Thermodynamics of Chemical Processes

Gareth Price

School of Chemistry, University of Bath

D0209936

Series sponsor: **ZENECA**

ZENECA is a major international company active in four main areas of business: Pharmaceuticals, Agrochemicals and Seeds, Specialty Chemicals, and Biological Products.

ZENECA's skill and innovative ideas in organic chemistry and bioscience create products and services which improve the world's health, nutrition, environment, and quality of life. ZENECA is committed to the support of education in chemistry and chemical engineering.

OXFORD UNIVERSITY PRESS
1998

Oxford University Press, Great Clarendon Street, Oxford OX2 6DP

Oxford   New York
Athens   Auckland   Bangkok   Bogota   Bombay
Buenos Aires   Calcutta   Cape Town   Dar es Salaam
Delhi   Florence   Hong Kong   Istanbul   Karachi
Kuala Lumpur   Madras   Madrid   Melbourne
Mexico City   Nairobi   Paris   Singapore
Taipei   Tokyo   Toronto   Warsaw

and associated companies in
Berlin   Ibadan

Oxford is a trade mark of Oxford University Press

Published in the United States
by Oxford University Press Inc., New York

A catalogue record for this book is available from the British Library

Library of Congress Cataloging in Publication Data
Price, Gareth J.
Thermodynamics of chemical processes/Gareth Price.
(Oxford chemistry primers; 56)
Includes bibliographical references and index.
1. Thermodynamics. I. Title. II. Series.
QD504.P75   1998   541.3'69–dc21   97-35358

ISBN 0 19 855963 1

Typeset by EXPO Holdings, Malaysia

Printed in Great Britain by
The Bath Press Ltd., Bath, Avon

Series Editor's Foreword

Oxford Chemistry Primers are designed to provide clear and concise introductions to a wide range of topics that may be encountered by chemistry students as they progress from the freshman stage through to graduation. The Physical Chemistry series will contain books easily recognized as relating to established fundamental core material that all chemists will need to know, as well as books reflecting new directions and research trends in the subject, thereby anticipating (and perhaps encouraging) the evolution of modern undergraduate courses.

In this Physical Chemistry Primer Gareth Price presents a clearly written and elegant introductory account of *Chemical Thermodynamics*. The book explains in simple terms the basic ideas and applications of a subject which is essential knowledge for any practising chemist. This Primer will be of interest to all students of chemistry (and their mentors).

Richard G. Compton
Physical and Theoretical Chemistry Laboratory, University of Oxford

Preface

Thermodynamics is a subject that is fundamental to the whole of chemistry but one that, at least at the outset, is amongst the least popular with students. It can be treated on a number of levels. First, there is the ultra-rigorous and pedantic approach which insists on deriving each equation and achieving a complete understanding of the minutiae. While this may be laudable it is likely to deter many students, particularly those with little background in physics and mathematics.

On the other hand, thermodynamics can be treated almost completely in terms of hand-waving arguments with barely an equation in sight, usually in a perceived response to the desires and requirements of students. Again, I feel that this is a mistake.

In this primer, I hope that I tread the middle ground between these two extremes. My overall aim is to introduce the concepts and ideas of how we treat energy changes during chemical processes and reactions and how we can use these ideas to correlate and explain chemical reactivity. I have not tried to be totally rigorous but hope that after reading and working through the book, students will be able to perform straightforward calculations and predictions on the course of reactions and the phase behaviour of compounds. In some places, non-IUPAC symbols and names have been used in the search for simplicity but I hope any deviations will be so obvious that no problems will be caused. No mathematical knowledge beyond that typically given at 'A-level' or equivalent is required. Some basic mathematical notes are provided for students without this level of background.

The book is based on a lecture course I have been delivering at the University of Bath now for some years. It has metamorphosed from earlier courses given by Peter Tiley and Tony Ashworth and I should acknowledge that base with gratitude. I am also grateful to Lee Clarke, one of my undergraduate tutorial students for his valuable comments and for trying the problems.

Above all, I hope that students and their teachers (at whatever level) will find the book useful.

Gareth Price
University of Bath, 1997.

Contents

1 Preamble: energy in chemical systems

Chemistry is essentially the science concerned with explaining molecular behaviour. Synthetic chemists try to change one type of molecule into another; structural chemists try to understand how atoms and molecules are arranged; physical and theoretical chemists try to explain how and why molecules behave in the way they do. Of course, these are generalities and most chemists are interested in more than one of these categories. This primer is concerned with the factors which govern chemical change. After studying the material covered, you should be able to make predictions about the behaviour of reactions and processes from first principles.

Thermodynamics is concerned with the study of *macroscopic* systems. We will be concerned with the properties of large, measurable amounts of matter rather than those of individual molecules. Indeed, many of the concepts covered in this primer do not need us to understand the molecular nature of matter or even to know that molecules exist. However, many topics are more easily understood (especially those concerned with entropy changes) if we consider what state the molecules are in and how they are arranged in our systems. Indeed, a related branch of physical chemistry—*molecular* or *statistical thermodynamics*—begins with the properties of individual molecules and derives the behaviour of chemical systems from these. However, this is outside the scope of this introductory primer.

As we develop the tools of thermodynamics we will find that one factor dominates our discussions. This is the *energy* of the system. In several universities I have visited, the material covered in this primer is actually titled '*energy changes in chemical processes*' to emphasize the usefulness of thermodynamics in chemistry. This chapter will focus on the origin of energy in chemical systems and particularly how it changes during chemical processes. Before proceeding to a detailed discussion of energy in chemical processes and reactions, there are some basic ideas common to the whole story that need to be established.

1.1 Chemical systems

In beginning a discussion of thermodynamics, we must be careful to define in a very precise manner, the part of the universe with which we are concerned. We describe this as the *system* of interest. The rest of the universe is described as the *surroundings*. Various kinds of system exist. *Isolated* systems occur where there can be no exchange of heat or matter with the surroundings. A *closed* system contains a fixed amount of matter but allows exchange of heat. *Adiabatic* systems allow no exchange of energy. Finally, the most common type of system which we encounter is an *open* system, in which both matter

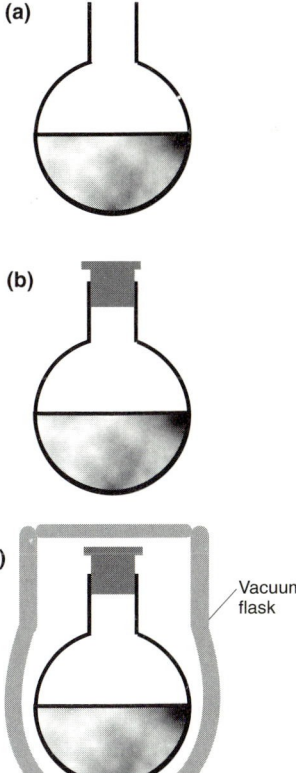

Fig. 1.1 Types of chemical systems:

(a) an open system;
(b) a closed system; and
(c) an isolated system.

In (c), the contents are held in the flask by the stopper and cannot be removed or added to. The vacuum flask prevents heat (energy) exchange with the surroundings. In (b), heat exchange can take place although the contents of the flask are fixed. In (a), exchange of both heat and matter with the surroundings is possible.

and energy can be exchanged with the surroundings. For example, a reaction taking place in a sealed, well-insulated calorimeter would be an isolated system, while the same reaction in an ordinary beaker would be an open system, as shown schematically in Fig. 1.1.

Changes in systems

Another factor requiring great care when discussing thermodynamics is the precise use of symbols and terminology. It is customary to represent the change in any property of a system by the symbol Δ. In order to be consistent, if we are interested in some property X, then Δ is precisely defined by

$$\Delta X = X \text{ after the change} - X \text{ before the change} = X_{\text{final}} - X_{\text{initial}} \qquad (1.1)$$

Thus, for a system being heated from a temperature T_1 to T_2, $\Delta T = T_2 - T_1$. For a chemical reaction, ΔX would be calculated as $X(\text{products}) - X(\text{reactants})$.

Ideal gases and atmospheric pressure

The concept of an *ideal gas* should be a familiar one. It is one of a number of so-called 'ideal' reference states which obey relatively simple laws and enable us to develop theoretical models of their behaviour. Another which we will meet later is that of an *ideal solution* (see Section 5.3).

In its most simple definition, an ideal gas is one whose properties obey the following simple equation:

$$pV = nRT \qquad (1.2)$$

where p is the pressure when n moles of a gas occupy a volume V at temperature T (measured in Kelvin). R is the *gas constant*, which is a fundamental constant and has the value (in SI units) of 8.314 J K^{-1} mol^{-1}. It can be shown that eqn 1.2 follows if we make some simple assumptions about the gas molecules. For example, the molecules are treated as point masses which have negligible size and no interaction with other molecules. Also, all collisions are elastic so that there is no energy transfer.

Although ideal reference states often have little to do with reality, the ideal gas is useful since most gases act ideally under the conditions in which we most frequently encounter them. It is only at very high pressures that significant deviations from eqn 1.2 occur. Indeed, many of the concepts of thermodynamics are most easily understood by considering an ideal gas and then relaxing the assumptions made to account for the behaviour of real systems. This is an approach which we will use in a number of places in this primer.

While discussing gases, it is appropriate to make some comment on the units used to measure pressures. Historically, pressures were measured by determining the height of a column of mercury that the pressure could support (Fig. 1.2). Thus, a common unit of 'pressure' was the 'millimetre of mercury' or 'mmHg', named the *torr* under the SI system. The pressure exerted by the atmosphere was about 730–780 mmHg depending on the weather. To standardize this and to provide a convenient unit for larger pressures, the *standard atmosphere*, 1 atm, was defined to be 760 torr. Even though we no longer rely on mercury manometers for pressure measurement (not the least for safety reasons) these units are still in widespread use.

Note: Δ is the upper case Greek letter delta and usually means an observable change to a system. In later discussions, we will use the lower case delta, δ, to signify a very small or *infinitesimally small* change.

Note: As an example, if a system is heated from 20 to 100 °C, ΔT would be $(100 - 20) = 80$ °C. If it were cooled from 20 to 0 °C, ΔT would be $(0 - 20) = -20$ °C.

Note: In the majority of thermodynamic work, we use the absolute temperature scale so that values are measured in Kelvin. The relationship needed to convert from degrees Celsius is:

$$T / K = T / °C + 273.15.$$

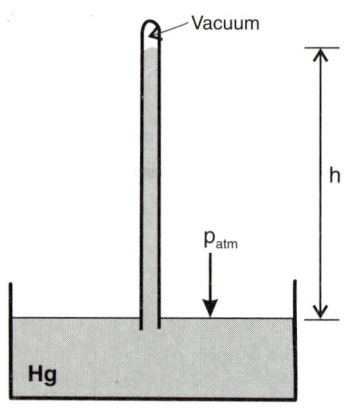

Fig. 1.2 Measuring pressure with a mercury barometer. The atmospheric pressure is equivalent to the height, *h*, of the mercury column it will support.

The SI unit of pressure is the pascal, Pa, which is the pressure when a force of 1 N acts on an area of 1 m². This is a rather small value—atmospheric pressure is around 100 000 Pa—so that for convenience we define another unit, the *bar*, where 1 bar = 10^5 Pa. The relationship between the various pressure units is shown in Table 1.1.

Table 1.1 Pressure units and conversion factors.

1 Pa	= 1 N m^{-2}
1 bar	= 10^5 Pa
1 torr	= 133.32 Pa
1 atm	= 1.013 bar
	= 101325 Pa
	= 760 Torr

Example 1.1 Calculate the volume occupied by 1 mole of an ideal gas at 25 °C and 1 atmosphere pressure.

From eqn 1.2, $pV = nRT$

$$(101\ 325\ \text{N m}^{-2})\ (V/\text{m}^3) = (1\ \text{mol})\ (8.314\ \text{J K}^{-1}\ \text{mol}^{-1})\ (298.15\ \text{K})$$

$$V = 0.02447\ \text{m}^3 = 24470\ \text{cm}^3$$

The last equality follows since 1 m³ = 10^6 cm³. Note that it is good practice in problem solving to include the units alongside the quantities in each equation. It can help to prevent errors.

1.2 Energy in chemistry

In everyday, common-sense terms, most people have a feel for what is meant by energy. We say that active people are 'full of energy'; we consider the energy usage in heating (or cooling) our homes. In a scientific sense, energy is more difficult to define, largely because it is a concept that occurs in many disguises. The interconversion of various types of energy is also familiar to us. The chemical energy in a fossil fuel such as coal or oil is released by burning and used to heat water to steam in a power station. This is used to produce motion in a turbine which converts mechanical energy to electrical energy which is delivered to our homes. The electrical energy can produce motion in an electric motor or heat in an electric fire or light from a lamp.

In physics textbooks, energy is often defined as *the capacity to do work*. This is a rather useless definition for us since we now have to define *work*. The easiest way to do this is by saying that work is done as a result of motion or a mechanical change. From Newton's laws, it is fundamental that these changes are the direct result of the action of a force. This leads us to define a unit for measuring energy, the joule, J. One joule is the work done when a force of 1 newton acts over a distance of 1 m. Thus,

$$1\ \text{J} = 1\ \text{N m} = 1\ (\text{kg m s}^{-2})\ \text{m} = 1\ \text{kg m}^2\ \text{s}^{-2}$$

Example 1.2 Calculate the energy needed to lift a book weighing 1 kg from the floor to a bookshelf 2 m high.

Here the force is provided by overcoming the action of gravity on the book.

Energy = force × distance
    = (1 kg × 9.8 m s^{-2}) × (2 m) = 19.6 kg m² s^{-2}
    = 19.6 J

A different type of energy with which we must be concerned in chemistry is *heat*. This is the form of energy that can be transferred as a result of temperature changes. Historically, this led to an alternative unit for energy—the calorie, cal. One calorie is the energy needed at 25 °C to heat 1 g of pure water by 1 °C.

To be consistent with the unit for work, the SI unit of energy is the joule and this will be used throughout this book. The calorie, however, remains in widespread use, particularly in US texts.

Quantifying heat changes

In order to quantify heat changes in systems, we introduce the heat capacity, c. The *molar heat capacity* is the amount of energy needed to raise the temperature of 1 mole of a substance by 1 K. The *specific heat capacity* is defined in an analogous way except that it refers to 1 gram of a substance. c therefore has units of J K^{-1} mol^{-1} or J K^{-1} g^{-1}. Hence, if q joules of heat are added to a system containing n mol or m grams of a substance and the temperature is raised from T_1 to T_2 then,

$$q = m\,c_s\Delta T = n\,c_m\Delta T \tag{1.3}$$

where $\Delta T = T_2 - T_1$. Here, the molar and specific heat capacities are denoted as c_m and c_s, although rather confusingly in many books the same symbol tends to be used for both. However, which one is being used should be readily apparent from its units. The two values are, of course, related by the molar mass of the compound.

Example 1.3   The heat capacity of liquid water is 4.18 J K^{-1}g^{-1}. Calculate the energy required to heat 1 mole of water from 25 to 90 °C.

Since the value is quoted per gram, it must be the specific heat capacity. Taking the molar mass of water as 18 g mol^{-1}, we can use eqn. 1.3

$$q/\text{J} = (1\,\text{mol} \times 18\,\text{g mol}^{-1})(4.18\,\text{J K}^{-1}\,\text{g}^{-1})(363.15 - 298.15\,\text{K})$$
$$= 4890.6\,\text{J} = 4.89\,\text{kJ}$$

Note that the temperatures were converted from degrees Celsius to kelvin. Even though this was not necessary in this particular problem, it is a good habit to always use K as the units for temperature.

Later, we will see that the value of the heat capacity will vary depending on the conditions under which it is measured and it also depends on temperature. However, this will suffice for present purposes.

Quantifying changes of work

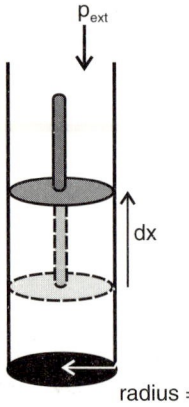

Fig. 1.3. Schematic diagram of a frictionless piston expanding against an external pressure.

We now need to be able to account for the energy encountered in chemistry as *work*. A simple example is to consider a reaction which forms a gas taking place inside a cylinder as shown in Fig. 1.3. The production of gas will push out the piston. Of course, in our thermodynamically ideal world, there is no friction and hence no heat caused as a result of the motion. The reaction can therefore be performed isothermally. The gas in the cylinder expands so that the piston moves a distance dx against the external pressure, p_{ext}, changing the volume from $V_{initial}$ to V_{final}.

The force acting on the piston can be calculated since

$$\text{pressure} = \frac{\text{force}}{\text{area}} \quad \text{or} \quad F = p_{ext}\pi r^2$$

hence

$$\text{work done} = \text{force} \times \text{distance moved}$$

$$dw = p_{ext}\,\pi r^2(dx) = -p_{ext}\,dV$$

$$w = \int_{V_{initial}}^{V_{final}} -p_{ext}\,dV \qquad (1.4)$$

dV is the change in volume of the gas on expansion. The negative sign arises since the movement is in the opposite direction to that in which the force is acting. dw is therefore the work done when the gas expands through a volume dV against a pressure, p_{ext}. Since the volume is continually changing, to find the total work done during the expansion, w, we need to integrate the expression between the initial and final volumes as shown in eqn 1.4.

Let us now think about how we might carry out the expansion. To simplify the problem we will consider *isothermal* changes, *i.e.* those at constant temperature. We start with the system at equilibrium and the piston at rest, with the pressures inside and outside the cylinder equalized.

Expansion against a constant pressure. Firstly, we could push the piston in to compress the gas and then release it. The expansion would then take place against a constant pressure, p_{ext}. In this case, eqn. 1.4 reduces to

$$w = -p_{ext} \int_{V_{initial}}^{V_{final}} dV = -p_{ext}(V_{final} - V_{initial})$$

or

$$w = -p_{ext}\,\Delta V \qquad (1.5)$$

Mathematical note: The volume, V, of a cylinder of radius r and height h is given by $V = (\pi r^2 h)$ or as (area) × (height).

Mathematical note: We use the notation dX to denote a small change in X, in fact *an infinitesimally small change*. This allows us to use the powerful tools of calculus to analyse the situation. Integration can be thought of as a method for summing a large number of small changes. The values above and below the integral sign are the limits of the summation, i.e. the final and initial values. So eqn 1.4 effectively sums a large number of $(p \times V)$ values between $V_{initial}$ and V_{final}.

Mathematical note: We now need to do some integration. In general, for values of $n \neq -1$, then

$$\int x^n dx = \frac{x^{n+1}}{n+1}$$

In other words, we increase the index (n) by 1 and divide by the new index. Other integrations will be introduced as needed. The integration involved here is

$$\int dx = \int x^0 dx = \frac{x^1}{1} = x$$

> *Example 1.4* Calculate the work done when 1 mole of a gas expands from 5 dm^3 to 10 dm^3 against a constant pressure of 1 atmosphere.
>
> ---
>
> $$\text{work} = -p_{ext}\,\Delta V = -(101\,325\text{ N m}^{-2})[(10-5)\text{ dm}^3 \times 10^{-3}\text{ m}^3\text{ dm}^{-3}]$$
> $$= -506.6\text{ J}$$
>
> Note that the work is negative since it is done in opposition to the external pressure.

Reversible expansion of an ideal gas. A second way of performing the experiment would be to push the piston in by rapidly increasing p_{ext} and then allowing the gas in the cylinder to expand slowly by gradually reducing the push on the piston or the external pressure. If the expansion were carried out very slowly, then the external pressure would only be very slightly lower than that inside the cylinder, p_{in}, at all stages of the expansion. As a thought experiment in our thermodynamic world, we could perform the expansion

Note: You may be wondering why we are indulging in this rather unrealistic exercise. It turns out that, with some of the topics to be discussed later, carrying them out in a reversible manner allows us to considerably simplify the mathematics. Real situations can then be considered in terms of reversible ones.

infinitely slowly where, effectively, we reduce the external pressure by a minute amount and allow the system to come back to equilibrium by equalizing the pressures. This is then repeated a large number of times until we reach the desired final pressure. By this means, we can assume that $p_{in} = p_{ext}$ throughout the experiment so that the system is always very close to equilibrium with its surroundings.

Such a process is called a *reversible change*. One of the criteria of reversibility is that the change can be reversed by an infinitesimal change in the opposite direction. If we simply modify the conditions so that p_{ext} becomes infinitesimally larger than p_{in}, the expansion would stop and be reversed

In the piston containing an ideal gas introduced in the previous section, we cannot use the simplification of eqn 1.4 since p_{ext} is continually changing. However, under reversible conditions, we can substitute $p_{ext} = p_{in}$ and use the ideal gas equation to describe the pressure.

Mathematical note: Eqn 1.6 arises from a rather special case of integration since

$$\int \frac{\mathrm{d}x}{x} = \ln x$$

and $(\ln a) - (\ln b) = \ln(a/b)$

$$w = \int_{V_{initial}}^{V_{final}} -p\mathrm{d}V = \int_{V_{initial}}^{V_{final}} -\frac{nRT}{V}\, \mathrm{d}V = nRT \int_{V_{initial}}^{V_{final}} \frac{1}{V}\mathrm{d}V$$

$$w_{rev} = -nRT\, ln\left(\frac{V_{final}}{V_{initial}}\right) \tag{1.6}$$

Mathematical note: An alternative way of interpreting the integral in eqn 1.4 is to consider it graphically. The value to be integrated is the product of pressure x volume and so can be thought of as finding the area of a plot of p versus V between the final and initial states. For an ideal gas, this can be represented as:

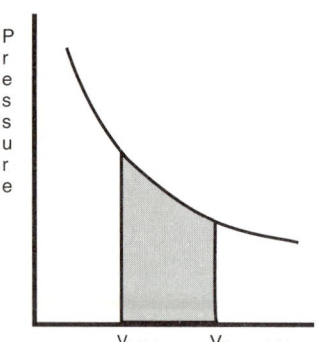

The shaded area is equivalent to the integral in eqn 1.4.

Example 1.5   Calculate the work done when 1 mole of a gas reversibly expands from 5 dm³ to 10 dm³ at 25 °C.

---

work $= -(1\text{ mol}) (8.314\text{ J K}^{-1}\text{ mol}^{-1})(298.15\text{ K})\, ln\, (10\text{ dm}^3/5\text{ dm}^3)$

$\quad\quad = -2478\text{ J} \times \ln (2)$

$\quad\quad = -1718\text{ J}$

The results from Examples 1.3 and 1.4 indicate that more work is done when the expansion occurs reversibly than when it occurs against a constant pressure. In fact, although we will not prove the fact here, the work obtained when a change is carried out reversibly is the *maximum* work available.

In chemical reactions involving gases, the volume changes can be large so that significant amounts of work can be done. In general, volume changes in systems containing only liquids or solids are much smaller and, as a first approximation, can be ignored.

Although discussion here has centred on heat and work, there are of course other sorts of energy change that can occur in chemical systems. One which will not concern us here is that due to mass changes in nuclear systems. The energy change here is governed by the Einstein equation and given by $\Delta E = \Delta mc^2$, where Δm is the loss in mass when nuclei are transformed and c is the speed of light. In the kind of chemical reactions usually considered, the mass does not change, so that we can safely ignore the nuclear energy. Similarly, we will not discuss the energy of electromagnetic radiation involved in photochemical processes

A form of energy which is rather more commonly met is in the form of electrical work that is done when electrons move around a system owing to a potential difference. If a charge Q moves through a potential difference E, then the work done is given by

$$w_{elec} = \int_{Q_{initial}}^{Q_{final}} -E \, dQ \qquad (1.7)$$

We will return to a discussion of the thermodynamics of electrochemical systems later in Section 4.10.

An accounting system for energy changes

At this stage, we need to have a scheme for accounting for energy changes. The one normally used is rather analogous to that used in keeping track of your personal finances in your bank account. (Your account is analogous to the chemical system.) Money given to you or which is added to your account is regarded as positive; money you spend or is charged to your account is negative. Similarly, for chemical systems:

> *Energy added to or work done on a system is POSITIVE.*
> *Energy removed from or work done by a system is NEGATIVE.*

Clearly, if a system gains an amount of energy q from the surroundings, then the change is $+q$ for the system and $-q$ for the surroundings. In Examples 1.2 and 1.3 the system was undergoing an expansion. Hence the system was doing work on the surroundings so that w was negative.

1.3   Internal energy and the First Law

In all matter, atoms and molecules are moving and so have energy. This is easy to see in fluids where the species undergo rotation and translational motion, giving rise to macroscopic effects such as gas pressure, Brownian motion, and diffusion. In other than monoatomic species, the chemical bonds are vibrating. Even in solids, the ions or molecules are vibrating (except at absolute zero temperature). This implies that they have energy. In addition, molecules possess potential energy as a result of their chemical bonds and the intermolecular forces. We call the sum of the kinetic and potential energy the *internal energy*, given the symbol U. In practice, to evaluate this energy, we would have to sum all the various contributions to the energy, which cannot easily be done. Good estimates of these parameters can be obtained from statistical thermodynamics but discussion of this is outside the scope of this primer. Rather than absolute values, however, it is more straightforward and useful to think of *changes* to this internal energy.

If we heat a gas, the molecules move faster, i.e. have more energy. If we compress a gas, as discussed above, its pressure will rise. From the definition of work given earlier, this means that the gas has a greater capacity to move the piston, i.e. it has greater energy.

We can summarize this in the *First Law of thermodynamics*. For an isolated system, the change in internal energy is the sum of heat, q, and work, w, changes.

$$\Delta U = q + w \qquad (1.8)$$

Note: Eqn 1.8 is more correctly written in the calculus form as

$$dU = dq + dw$$

In more advanced work this allows calculus to be used in deriving a number of useful relationships.

There are a number of ways of expressing this law. One is that the total energy of an isolated system is constant. Perhaps the most common statement is the principle of conservation of energy: *'Matter can neither be created or destroyed, merely interconverted between forms'.* This, of course, discounts any nuclear energy changes which were referred to earlier.

Example 1.6   The temperature of 1 mol of a substance is raised by heating it with 750 J of energy. It expands and does the equivalent of 200 J of work. Calculate the change in internal energy.

Since the temperature increases, $q = +750$ J. In an expansion, the system does work so that $w = -200$ J. Using eqn. 1.7

$$\Delta U = q + w = (+750) + (-200) = +550 \text{ J}$$

Perhaps a more useful way of thinking of the internal energy is to write eqn 1.8 in the form

$$\Delta U = q + p\Delta V$$

If we now heat a system which cannot do p–V type work, i.e. at fixed volume, then

Mathematical note: The style $(X)_{p,T}$ is shorthand for '*The value of X at constant p and T*'. The value inside the parentheses is evaluated at constant values of the subscripts.

$$\Delta U = q + 0 = q_{\mathrm{v}} \tag{1.9}$$

where q_{v} indicates a heat change at *constant volume*. In other words, the change in internal energy can be thought of as the heat change for a process carried out at constant volume.

Equation 1.9 is completely general. However, an isothermal change to the internal energy of ideal gases produces a rather specialized result which is worthy of further comment. Since one of the assumptions is that the molecules are far apart and non-interacting, the internal energy depends only on the temperature and not on the volume which contains the gas or the pressure it exerts. Thus, if we compress the ideal gas *isothermally* at constant temperature, there must be a net outflow of heat from the system. Alternatively, an expansion must be accompanied by the absorption of heat.

Mathematical note: The formalism $\frac{\mathrm{d}X}{\mathrm{d}T}$ comes from *differential calculus*. Differentiation can be thought of as the opposite of integration. In particular, the differential tells us how the value of X changes for a small change in T. It is the equivalent of measuring the slope of a graph of X versus T, and is particularly useful where the slope continuously changes its value.

$$\Delta U = 0 = q + w \quad \therefore q = -w$$

Heat capacities of gases

We can correlate the amount of energy required for the heating a substance by using an equation analogous to eqn 1.3. If an amount of energy $\mathrm{d}q_v$ added to a gas at constant volume causes a temperature change $\mathrm{d}T$, then the *constant volume molar heat capacity*, c_{v}, is given by

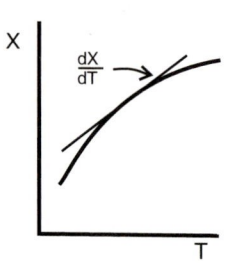

$$\mathrm{d}q_{\mathrm{v}} = c_{\mathrm{v}}\mathrm{d}T \quad \text{or} \quad c_{\mathrm{v}} = \left(\frac{\mathrm{d}q}{\mathrm{d}T}\right)_{\mathrm{v}} \tag{1.10}$$

Note: In the case of heat capacities, the variation of q with T is much closer to a straight line than the curve in this figure.

the subscript 'v' indicating that the heating is carried out at constant volume. The differential form of eqn 1.10 is more strictly correct since c_{v} in fact changes somewhat with temperature. However, for small temperature ranges the variation is small and heat capacities are often considered constant. We will see in the next chapter how to deal with situations where this assumption cannot be made.

State functions

The internal energy is an example of a *state function*. Its value depends only on the energy of the system, defined by the pressure and temperature (or volume and temperature) of the system, i.e. the present state of the system. There is no dependence on previous behaviour of the system nor of how the system achieved its present state.

State functions have a series of properties, the most important of which is that the change in value of the function depends only on the final and initial conditions and is independent of the path between them. An analogy from everyday life (and hence inexact!) is in climbing a mountain. You could climb up the hill in a straight line to reach the summit. Alternatively, you could wander around the mountain in a gentle climb and eventually reach the summit. The distance that you cover in each case will be very different; the change in altitude will, however, be the same. The altitude is therefore analogous to a state function. A number of other state functions will be introduced as we progress through the discussion of thermodynamics.

1.4 Measurement of internal energy changes

From the above discussion, the change in internal energy during a reaction is simply the heat change during the reaction if it is conducted at constant volume. Thus, carrying out a reaction under conditions where the volume is fixed provides the easiest way of measuring ΔU. The relation between ΔU and other measurable parameters will be discussed in the following chapter.

As an example, ΔU during a combustion reaction can be measured using a *bomb calorimeter*, shown schematically in Fig. 1.4. The reaction chamber or (totally misnamed!) *bomb* which contains the substance under study is constructed from thick (~1 cm) stainless steel, ensuring that the system volume does not change. After pressurizing with oxygen gas, the reaction can be started by electrically heating a piece of wire in contact with the reactants.

The heat given out on combustion warms up both the metal of the calorimeter and the water that surrounds it. The temperature is measured precisely to ± 0.001 °C with an accurate thermometer. We can calculate the effective heat capacity, c, of the calorimeter since

$$c_{\text{calorimeter}} = m_{\text{water}} \, c_{\text{water}} + m_{\text{steel}} \, c_{\text{steel}} \tag{1.11}$$

where m represents the mass of water or steel and c the appropriate specific heat capacity. If the combustion of n moles of a compound results in a temperature increase ΔT, then the internal energy change is

$$\Delta U^{\text{combustion}} = -c_{\text{calorimeter}} \Delta T / n \tag{1.12}$$

the negative sign arises from our sign convention, since the heat is given out from the system.

In practice, rather than using eqn 1.11, $c_{\text{calorimeter}}$ is determined by a calibration. The temperature rise when a known mass of a compound with accurately known $\Delta U^{\text{combustion}}$ undergoes reaction is measured and $c_{\text{calorimeter}}$ calculated from eqn 1.12. Naphthalene or benzoic acid are commonly used for this purpose. Alternatively, an electrical calibration can be performed. If a current I passes through the wire at a voltage V for time t, the energy delivered is given by ($I \times V \times t$). Measurement of the temperature increase caused by

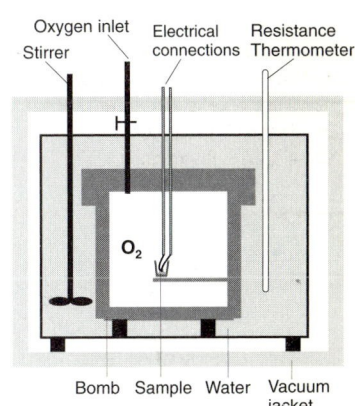

Oxygen inlet Electrical Resistance
Stirrer   connections Thermometer

O_2

Bomb Sample Water Vacuum
                      jacket

Figure 1.4 Schematic diagram of a bomb calorimeter for measuring changes in internal energy on combustion.

Note: The chemical changes involved in the reactions for calibrating the calorimeter are:

$$C_{10}H_{8(s)} + 12O_{2(g)} \rightarrow 10CO_{2(g)} + 4H_2O_{(l)}$$
$$C_6H_5COOH_{(s)} + 7\tfrac{1}{2}O_{2(g)}$$
$$\rightarrow 7\,CO_{2(g)} + 3H_2O_{(l)}$$

this energy applied to a wire inside the bomb also allows $c_{\text{calorimeter}}$ to be calculated, as in Example 1.7.

Example 1.7   The combustion of 0.6475 g of naphthalene in a bomb pressurized with oxygen at 25 °C resulted in a temperature increase of 2.424 °C. Under the same conditions, passage of an electric current of 1.5 A at 15 V for 15 min raised the temperature by 1.890 °C. Calculate $\Delta U^{\text{combustion}}$ for naphthalene.

First, we need to find the heat capacity of the calorimeter.

Energy supplied $= IVt = 1.5 \text{ A} \times 15 \text{ V} \times (15 \times 60) \text{ s} = 20\,250 \text{ J}$

$$c_{\text{calorimeter}} = \frac{\text{energy supplied}}{\text{temperature change}} = \frac{20\,250 \text{ J}}{1.89 \text{ K}} = 10\,714 \text{ JK}^{-1}$$

Now, turning to the data for naphthalene, $0.6475 \text{ g} = 5.05 \times 10^{-3}$ moles. Therefore,

$$5.05 \times 10^{-3} \text{ mol releases } c_{\text{calorimeter}} \Delta T = 10\,714 \text{ JK}^{-1} \times 2.424 \text{ K}$$
$$= 25\,967 \text{ J}$$

$$\Delta U^{\text{combustion}} = -25\,967 \text{ J}/5.05 \times 10^{-3} \text{ mol} = -5\,124\,040 \text{ J mol}^{-1}$$

Hence,

$$\Delta U^{\text{combustion}} \text{ for naphthalene at 25 °C} = -5142 \text{ kJ mol}^{-1}$$

Note that energy is evolved from the naphthalene to raise the temperature of the water. The change in internal energy of the system is therefore *negative*.

1.5   Concluding remarks

The discussion in this chapter has centred on the internal energy of a system and allowed us to introduce a number of useful concepts. However, ΔU is of limited use since few chemical processes are conducted at constant volume. With these basic tools though, we can now begin to tackle some more typical chemical processes. In many cases, we will investigate in detail the situation for a model system, for example an ideal gas, and then generalize the equations developed to describe more realistic situations.

2 Enthalpy and thermochemistry

In the previous chapter, we discussed heat changes for processes conducted under constant volume conditions where no pressure–volume work was done by the system. However, the majority of reactions and processes that we encounter will be conducted at constant pressure, e.g. in open vessels in the laboratory. Thus, we need a function to describe energy changes at constant pressure. This is the enthalpy, given the symbol H. An old fashioned name for the enthalpy is the 'heat content' of a compound, although we will usually be more interested in enthalpy *changes* during chemical processes. The importance of measuring enthalpy changes to give fundamental information on reactions will soon become apparent. However, to a practical chemist a knowledge of how much heat a reaction needs—or how much it evolves—can be much more significant (and in the latter case, a matter of life and death on a large scale!).

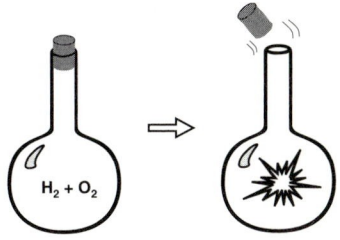

Fig. 2.1 An example of a reaction involving changes of both heat and work.

The most straightforward (and hence least rigorous!) way of defining the enthalpy is to consider the two parts of the process. As shown schematically in Fig. 2.1, the system undergoes a heat change and it does work by moving the stopper of the flask against the atmosphere. We saw in the previous chapter that energy could be added to the system by heating it (i.e. ΔU) and by doing work on the system (given at constant pressure by $p\Delta V$). Thus, the change in energy at constant pressure, or *enthalpy change*, ΔH, is given by

$$\Delta H = q_p = \Delta U + p\Delta V \tag{2.1}$$

which leads to a formal definition of the enthalpy, H as

$$H = U + pV \tag{2.2}$$

It is important to have an appreciation of the magnitude of the difference between ΔU and ΔH. Considering eqn 2.1, the difference arises from the $p\Delta V$ term. In general, the volume changes involved in liquid or solid systems are very small and can be neglected with little error. Thus, $\Delta H \approx \Delta U$. However, when gases are involved, the volume changes can be large. If the gas acts ideally, then $p\Delta V = \Delta n\, RT$, where Δn is the change in the number of moles of gas. At 25 °C, 298.15 K, $(\Delta H - \Delta U) = 2.48$ kJ for each mole of gas involved; at 500 °C, $(\Delta H - \Delta U) = 6.43$ kJ mol^{-1}. In terms of the magnitude of enthalpy changes, these values are not negligible.

In considering enthalpy changes, we use the same sign convention as for internal energy. Processes resulting in an evolution of heat are *exothermic* and have negative ΔH; heat absorption corresponds to positive ΔH and arises from *endothermic* processes.

2.1 Enthalpy changes in chemical systems—Hess's law

Since the enthalpy is composed of a number of state functions, H must also be a state function and have the properties described in Section 1.3. One of the main practical consequences of this is that the enthalpy change for a reaction is merely the difference between the final state of the system (i.e. the products) and the initial state (i.e. the reactants). Any intermediate stages or reactions do not matter. This principle is embodied in Hess's law which can be stated as:

The total enthalpy change for a reaction is independent of the path by which the reaction occurs.

As a straightforward example, say we are interested in the oxidation of carbon monoxide to carbon dioxide at 1 bar and 25 °C. We could conduct the reaction directly and measure ΔH.

$$(1) \quad CO_{(g)} + \tfrac{1}{2}O_{2(g)} \rightarrow CO_{2(g)} \quad \Delta H_1 = -283.0 \text{ kJ mol}^{-1}$$

Alternatively, the reaction of graphite with oxygen could be carried out under two different stoichiometries.

$$(2) \quad C_{(graph.)} + \tfrac{1}{2}O_{2(g)} \rightarrow CO_{(g)} \quad \Delta H_2 = -110.5 \text{ kJ mol}^{-1}$$
$$(3) \quad C_{(graph.)} + O_{2(g)} \rightarrow CO_{2(g)} \quad \Delta H_3 = -393.5 \text{ kJ mol}^{-1}$$

Hess's law allows us to treat the chemical equations and the enthalpy changes in the same way as algebraic equations. Subtracting (2) from (3) gives the equivalent of (1) and leads to the same result for ΔH.

$$[C_{(graph.)} + O_{2(g)}] - [C_{(graph.)} + \tfrac{1}{2}O_{2(g)}] \rightarrow [CO_{2(g)}] - [CO_{(g)}]$$
$$CO_{(g)} + \tfrac{1}{2}O_{2(g)} \rightarrow CO_{2(g)}$$
$$\Delta H = (-393.5) - (-110.5) = -283.0 \text{ kJ mol}^{-1}$$

or
$$\Delta H_3 = \Delta H_1 + \Delta H_2$$

An alternative method of displaying these enthalpy changes and of performing the calculations is to use graphical diagrams of the energy levels. Such an approach is illustrated in Fig. 2.2.

This method has a number of important applications in chemistry. The most obvious consequence of Hess's law (and the First Law of Thermodynamics) is that, for a reaction, $\Delta H_{forward} = -\Delta H_{reverse}$. A more important use is the calculation of enthalpy changes for reactions which would be experimentally difficult or even impossible to carry out in practice, as in Example 2.1.

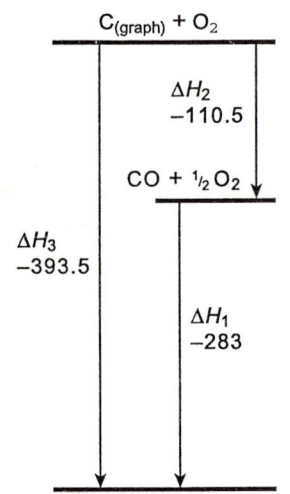

Fig. 2.2 Graphical display of the enthalpy of reaction for carbon and oxygen. Enthalpy changes in kJ mol^{-1}.

Example 2.1 The hydrocarbons ethene and ethane are by-products from the cracking of oil. The enthalpy change at 25 °C for the reaction of graphite and hydrogen gas to form each of the compounds is $+52.26$ kJ mol^{-1} and -84.68 kJ mol^{-1}, respectively. Calculate the enthalpy change for the hydrogenation of ethene to give ethane.

From the data given

Ethene : $2C_{(graph.)} + 2H_{2(g)} \rightarrow C_2H_{4(g)} \quad \Delta H = +52.26 \text{ kJ mol}^{-1}$ (A)
Ethane : $2C_{(graph.)} + 3H_{2(g)} \rightarrow C_2H_{6(g)} \quad \Delta H = -84.68 \text{ kJ mol}^{-1}$ (B)

The reaction of interest is
$$C_2H_{4(g)} + H_{2(g)} \rightarrow C_2H_{6(g)} \quad \Delta H = ?$$

If we subtract reaction A from reaction B,

$$[2C_{(graph)} + 3H_{2(g)}] - [2C_{(graph.)} + 2H_{2(g)}] \rightarrow [C_2H_{6(g)} - C_2H_{4(g)}]$$

We can now treat the chemical equations exactly like an algebraic equation

$$H_{2(g)} \rightarrow C_2H_{6(g)} - C_2H_{4(g)} \quad \Delta H = \Delta H(B) - \Delta H(A)$$

$$C_2H_{4(g)} + H_{2(g)} \rightarrow C_2H_{6(g)} \quad \Delta H = -84.68 - (+52.26) \text{ kJ mol}^{-1}$$

The enthalpy change for the hydrogenation is therefore -136.94 kJ mol^{-1}.

Note that ethene is sometimes called ethylene and is the base chemical for the widely used plastic polyethylene or polythene

Ethane   **Ethene**

$CH_3 - CH_3$ $CH_2 = CH_2$

$$\left\{ \begin{array}{c} CH_2 \\ \end{array} \begin{array}{c} CH_2 \\ CH_2 \end{array} \begin{array}{c} CH_2 \\ \end{array} \right\}_{10-10^6}$$

Poly(ethylene)

Note: Chemical structures of compounds in Example 2.1

Enthalpy changes in chemical reactions

As has already been stated, the enthalpy change for a reaction is simply the heat absorbed or evolved when the reaction occurs at constant pressure and is the difference in enthalpy between the products and the reactants. However, ΔH is dependent on the temperature and pressure at which the reaction occurs as well as the state of the components. Thus we need to define a set of standard conditions which can be used to report results. These are:

Standard temperature: 25 °C, 298.15 K
Standard pressure: 1 bar, 101 325 Pa
Standard state: pure component at 1 bar pressure

Note that these conditions are used as the standard for reporting thermodynamic data and are not to be confused with the 'STP' used for gas properties which is 1 atmosphere and 0 °C.

We can therefore define a *standard enthalpy of reaction* as the enthalpy change at 1 bar and 25 °C for a reaction with all components in their standard states. This is given the symbol ΔH°_{298}. The superscript $^\circ$ indicates that the components are in their standard states at standard pressure and the subscript indicates the temperature.

Using this gives us a convenient way of conveying information. For example, the following equation

$$N_{2(g)} + 3H_{2(g)} \rightarrow 2NH_{3(g)} \quad \Delta H^\circ_{298} = -92.2 \text{ kJ mol}^{-1}$$

tells us that, when 1 mol of nitrogen reacts with three moles hydrogen to form 2 moles of ammonia at 25 °C and 1 bar pressure, -92.2 kJ is evolved. Note that care must be taken when making statements such as 'the enthalpy of reaction is -92.2 kJ mol^{-1}'. This would refer to the above equation as written so 'per mole' refers to 'per mole of nitrogen consumed'. We could equally well write

$$\tfrac{1}{2}N_{2(g)} + 1\tfrac{1}{2}H_{2(g)} \rightarrow NH_{3(g)} \quad \Delta H^\circ_{298} = -46.1 \text{ kJ mol}^{-1}$$

which would be interpreted as the enthalpy change per mole of ammonia formed.

Standard enthalpy of formation

The formation of a compound from its elements as typified by the last example discussed—the so-called *Haber* process for the synthesis of ammonia—is an important class of reaction. We define the enthalpy change for this under standard conditions as the *standard enthalpy of formation* of the compound, given the symbol ΔH_f°.

ΔH_f° is the enthalpy change when 1 mole of a compound is formed under standard conditions from its constituent elements in their standard states.

This is an important fundamental property of a compound and is very useful in order to correlate energy changes during chemical reactions. In general, values are quoted at the standard temperature of 25 °C or 298.15 K. Tables of $\Delta H_{f,298}^\circ$ are available and a selection of values is given in Appendix 1.

It is not possible to define an absolute measure of enthalpy so that we need to have a reference value on which to base our values. Following on from the definition of $\Delta H_{f,298}^\circ$, it is convenient to assume that for elements in their standard states, $\Delta H_{f,298}^\circ$ must be zero, and this indeed is the reference state on which enthalpy changes are based. Thus, the standard enthalpy of formation of ammonia is the standard enthalpy change for the reaction

$$\tfrac{1}{2}N_{2(g)} + \tfrac{3}{2}H_{2(g)} \rightarrow NH_{3(g)} \quad \Delta H_{298}^\circ = \Delta H_{f,298}^\circ(NH_3) = -46.1 \text{ kJ mol}^{-1}$$

As usual in thermodynamics, the definition of ΔH_f° is a very precise statement and contains some traps for the unwary. The key phrase is *standard state*. Where several forms of an element exist, the one with lowest energy must be used as the standard state. A good example is carbon, where solid graphite is taken as standard in preference to diamond or fullerene. Other cases need more care. For example, bromine and iodine often react in the vapour phase. However, their standard states are the liquid and solid, respectively.

$$H_{2(g)} + \tfrac{1}{2}O_{2(g)} \rightarrow H_2O_{(l)} \quad H_{298}^\circ = H_{f,298}^\circ(H_2O_{(l)}) = -285.8 \text{ kJ mol}^{-1}$$
$$C_{(graph.)} + O_{2(g)} \rightarrow CO_{2(g)} \quad \Delta H_{298}^\circ = \Delta H_{f,298}^\circ(CO_{2(g)}) = -393.5 \text{ kJ mol}^{-1}$$
$$H_{2(g)} + Br_{2(l)} \rightarrow 2HBr_{(g)} \quad \Delta H_{298}^\circ = 2\Delta H_{f,298}^\circ(HBr_{(g)}) = -72.8 \text{ kJ mol}^{-1}$$

In each of the three reactions, the enthalpy of the reactants is zero so that the enthalpy change for the reaction represents the enthalpy of formation of the products.

Example 2.2   Calculate the enthalpy change for the gas phase reaction between hydrogen and iodine vapour at 25 °C.

$$H_{2(g)} + I_{2(g)} \rightarrow 2\,HI_{(g)}$$

$\Delta H_{f,298}^\circ$ for $H_{2(g)}$ is zero by definition since it is an element in its standard state. However, the standard state for iodine is the solid so that $\Delta H_{f,298}^\circ$ for $I_{2(g)} \neq 0$. $\Delta H_{f,298}^\circ$ for $HI_{(g)}$ is 26.48 kJ mol^{-1}.

$$\Delta H_{f,298}^\circ(I_{2(g)}) = \Delta H_{f,298}^\circ(I_{2(s)}) + \Delta H^\circ(I_{2(s)} \rightarrow I_{2(g)}) = 0 + 62.44 \text{ kJ mol}^{-1}$$

The enthalpy change for the reaction is therefore given by eqn 2.3

$$\Delta H^\circ = [2\,\Delta H_{f,298}^\circ(HI_{(g)})] - [\Delta H_{f,298}^\circ(H_{2(g)}) + \Delta H_{f,298}^\circ(I_{2(g)})]$$
$$= [2(26.48)] - [0 + 62.44]$$
$$= -9.48 \text{ kJ (mol } H_2)^{-1}$$

Note that because of the stoichiometry, $\Delta H^\circ = -4.74$ kJ (mol HI)$^{-1}$

Applications of enthalpies of formation

To illustrate the use of $\Delta H^\circ_{f,298}$ values, consider the gas phase reaction of acetylene (ethyne) to benzene.

$$3C_2H_{2(g)} \rightarrow C_6H_{6(g)}$$

We can use Hess's law to write this reaction as if it proceeded via an intermediate stage:

$$3C_2H_{2(g)} \xrightarrow{\Delta H^\circ_a} [6C_{(graph)} + 3H_{2(g)}] \xrightarrow{\Delta H^\circ_b} C_6H_{6(g)}$$

so that ΔH°_{298} (reaction) $= \Delta H^\circ_a + \Delta H^\circ_b$. Note that such a reaction sequence may not be possible in practice but this doesn't prevent us thinking about it in thermodynamic terms!

From the definitions given,

$$\Delta H^\circ_a = 3[-\Delta H^\circ_{f,298}(C_2H_{2(g)})] \quad \text{and} \quad \Delta H^\circ_b = \Delta H^\circ_{f,298}(C_6H_{6(g)}).$$

Thus,

$$\Delta H^\circ_{298}(\text{reaction}) = \{3[-\Delta H^\circ_{f,298}(C_2H_{2(g)})]\} + \{\Delta H^\circ_{f,298}(C_6H_{6(g)})\}$$

or

$$\Delta H^\circ_{298}(\text{reaction}) = \Delta H^\circ_{f,298}(C_6H_{6(g)}) - 3\Delta H^\circ_{f,298}(C_2H_{2(g)})$$

Using the data from Appendix 1,

$$\Delta H^\circ_{298}(\text{reaction}) = +82.9 - (3 \times +226.7) = -597.2 \text{ kJ mol}^{-1}$$

so the reaction is highly exothermic. The graphical calculation of reaction enthalpies is illustrated in schematic form in Fig. 2.3.

A more general way of writing the result of this calculation is

$$\Delta H^\circ_{298}(\text{reaction}) = \Delta H^\circ_{f,298}(\text{products}) - \Delta H^\circ_{f,298}(\text{reactants})$$

It can readily be shown that this equation will hold for any reaction so that we can formulate eqn 2.3 where ν_i represents the stoichiometric coefficient.

$$\Delta H^\circ_{298}(\text{reaction}) = \sum \nu_i \Delta H^\circ_{f,298}(\text{products}) - \sum \nu_i \Delta H^\circ_{f,298}(\text{reactants}) \quad (2.3)$$

Hence, for a general reaction of the form

$$\alpha A + \beta B \rightarrow \gamma C + \delta D \quad (2.4)$$

then

$$\Delta H^\circ_{298}(\text{react.}) = [\gamma \Delta H^\circ_{f,298}(C) + \delta \Delta H^\circ_{f,298}(D)] - [\alpha \Delta H^\circ_{f,298}(A) + \beta \Delta H^\circ_{f,298}(B)]$$

This equation can now be used to calculate the enthalpy change for any reaction where we have $\Delta H^\circ_{f,298}$ data for the components, for instance the reaction shown in Example 2.3. While the energy level diagram approach illustrated in Figs 2.2 and 2.3 works well in straightforward reactions such as those discussed so far, in more complicated cases it can be easy to make errors. The use of eqn 2.3 (several times if necessary) is more readily applied to all cases. Thus we are now in a position to use tabulated $\Delta H^\circ_{f,298}$ values to calculate the enthalpy of any reaction—but so far only at 25 °C. We will now proceed to see how we can calculate ΔH° under other conditions.

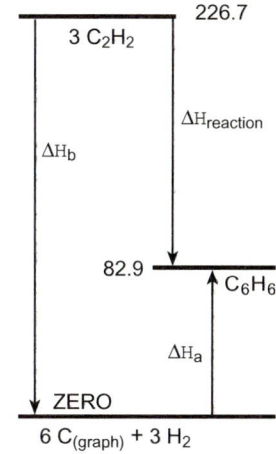

Fig. 2.3 Schematic calculation of $\Delta H^\circ_{298(\text{reaction})}$.

Mathematical note: The symbol Σ is shorthand for the *summation* of the terms which follow, e.g.

$$\sum\nolimits_1^5 (a_i) = a_1 + a_2 + a_3 + a_4 + a_5$$

Example 2.3 Calculate the enthalpy of reaction at 25 °C for the following reaction.

$$3Fe_2O_{3(s)} + 2NH_{3(g)} \rightarrow 6FeO_{(s)} + 3H_2O_{(l)} + N_{2(g)}$$

Using eqn 2.3 and data from Appendix 1, ΔH°_{298} is given by:

$$\Delta H^\circ_{298}(\text{reaction}) = \sum v_i \Delta H^\circ_{f,298}(\text{products}) - \sum v_i \Delta H^\circ_{f,298}(\text{reactants})$$

$$= [6\Delta H^\circ_{f,298}(FeO_{(s)}) + 3\Delta H^\circ_{f,298}(H_2O_{(l)}) + \Delta H^\circ_{f,298}(N_{2(g)})]$$
$$-[3\Delta H^\circ_{f,298}(Fe_2O_{3(s)}) + 2\Delta H^\circ_{f,298}(NH_{3(g)})]$$

$$= [6(-266.3) + 3(-285.8) + 0] - [3(-824.2) + 2(-46.1)]$$

$$= (-2455.2) - (-2564.8)$$

$$= 109.6 \text{ kJ mol}^{-1}.$$

Note that for clarity, the units have been omitted from the intermediate stages of the calculation. All enthalpy values are in kJ mol^{-1}.

2.2 Variation of enthalpy with temperature

The discussion in the previous sections allows us to calculate enthalpy changes at 25 °C. However, most reactions or processes will take place at other temperatures so we need a method for adapting our enthalpy values.

In an exactly analogous manner to the definition of c_v at constant volume, eqn 1.10, we can define a *constant pressure heat capacity*, c_p, as the amount of heat energy needed to raise the temperature of 1 mole of substance by 1 K at constant pressure.

$$c_p = \left(\frac{dq_p}{dT}\right)_p = \left(\frac{dH}{dT}\right)_p \quad \text{or} \quad dH = c_p dT \qquad (2.5)$$

Thus, the enthalpy of a substance must increase by an amount ($c_p \, dT$) if heated by an amount dT. It therefore follows that, if a substance is heated from T_1 to T_2, then the change of enthalpy is given by

Note: The definition of enthalpy allows us to derive a straightforward relation between the two heat capacities in the case of ideal gases.

Since, $H = U + pV$, we can write, for 1 mole of gas,

$$H = U + RT$$

For a small change of temperature,

$$dH = dU + R \, dT$$

$$\frac{dH}{dT} = \frac{dU}{dT} + R$$

$$c_p = c_v + R$$

or $\quad c_p - c_v = R$

with c_p and c_v defined on a molar basis.

$$\int dH = \int_{T_1}^{T_2} c_p dT \qquad (2.6)$$

This is a basic form of the *Kirchoff equation*. At first glance, this looks like a straightforward integral but, like c_v, c_p varies with temperature. As a first approximation, we could treat c_p as independent of temperature since the variation is small and is often insignificant for small temperature ranges. A somewhat better approach is to use the average value of c_p over the range involved. In this case, eqn 2.6 can be integrated to give

$$H_{T_2} - H_{T_1} = \bar{c}_p(T_2 - T_1) \qquad (2.7)$$

where $\bar{c}_p$ represents the *mean* heat capacity of the compound over the temperature range T_1 to T_2. A particularly useful form of this equation is to recall that standard enthalpies of formation are defined at 25 °C. Hence, we can write

$$\Delta H_{f,T}^{\circ} = \Delta H_{f,298}^{\circ} + \bar{c}_p(T - 298.15) \qquad (2.8)$$

A more accurate method for dealing with the temperature dependence of c_p will be given in the following section but we will first see how eqn 2.8 can be applied to enthalpies of reaction.

Enthalpies of reaction

Consider a model reaction where 1 mol of reactants, R, are converted into 1 mol of products, P.

$$R \rightarrow P$$

The enthalpy of R at 25 °C is $\Delta H_{f,298}^{\circ}$ (R). Using eqn 2.8, the enthalpy at some other temperature T is $[\Delta H_{f,298}^{\circ}$ (R) $+ \bar{c}_p(R) (T - 298.15)]$. The same relations hold for the products, P. Thus,

$$\Delta H_T^{\circ}(\text{reaction}) = \Delta H_{f,T}^{\circ}(P) - \Delta H_{f,T}^{\circ}(R)$$

$$= [\Delta H_{f,298}^{\circ}(P) + \bar{c}_p(P)(T - 298.15)] - [\Delta H_{f,298}^{\circ}(R) + \bar{c}_p(R)(T - 298.15)]$$

$$= [\Delta H_{f,298}^{\circ}(P) - \Delta H_{f,298}^{\circ}(R)] + [\bar{c}_p(P) - \bar{c}_p(R)](T - 298.15)\}$$

Hence, $$\Delta H_T^{\circ}(\text{reaction}) = \Delta H_{298}^{\circ}(\text{reaction}) + \Delta\bar{c}_p(T - 298.15) \qquad (2.9)$$

where we again use the formalism that $\Delta\bar{c}_p$ is the difference in mean heat capacities between the products and the reactants. For completion, we write

$$\Delta\bar{c}_p = \sum \nu_i \bar{c}_p(\text{products}) - \sum \nu_i \bar{c}_p(\text{reactants}) \qquad (2.10)$$

noting that the stoichiometry must also be taken into account.

Exercise. Derive eqn 2.9 in terms of the general reaction given in eqn 2.4 in order to satisfy yourself that it is indeed completely general.

Example 2.4   The complete combustion of ethane releases 1558.8 kJ mol^{-1} at 25 °C. Calculate ΔH°(combustion) at 100 °C.

$\bar{c}_p/J\ K^{-1}\ mol^{-1}$: $C_2H_{6(g)}$ 52.6 : $O_{2(g)}$ 29.4 : $CO_{2(g)}$ 37.1 : $H_2O_{(l)}$ 75.3

$$C_2H_{6(g)} + 3\tfrac{1}{2}O_{2(g)} \rightarrow 2CO_{2(g)} + 3H_2O_{(l)} \quad \Delta H_{298}^{\circ} = -1558.8\ \text{kJ mol}^{-1}$$

$$\Delta c_p = [2\bar{c}_p(CO_2) + 3\bar{c}_p(H_2O) - \bar{c}_p(C_2H_6) + 3\tfrac{1}{2}\bar{c}_p(O_2)$$
$$= [2(37.1) + 3(75.3) - [52.6 + 3\tfrac{1}{2}(29.4)]$$
$$= 144.6\ \text{J K}^{-1}\ \text{mol}^{-1}$$

$$\Delta H_{373}^{\circ} = \Delta H_{298}^{\circ} + \Delta\bar{c}_p\Delta T$$
$$= -1558.8 \times 10^3(\text{J mol}^{-1}) + 144.6\ (\text{J K}^{-1}\ \text{mol}^{-1})(75\ \text{K})$$
$$= -1547.9\ \text{kJ mol}^{-1}$$

A more accurate treatment of heat capacities

Mathematical note: It is common practice that, where the precise form of a function is unknown, a power series approach is used. The first term is a constant, the second adds a linear variation, the third introduces some curvature, the fourth a more complex curvature, and so on until the function fits the experimental data with sufficient accuracy for the purposes at hand.

So far, we have assumed that the use of mean heat capacities gives us sufficient accuracy. In fact, c_p can change significantly in some cases. For more precise work, or where large temperature ranges are involved, a more complete description of the heat capacity must be used. The variation with temperature is usually represented as a *power series*.

$$c_p(T) = a + bT + cT^2 + dT^3 + \ldots \tag{2.11}$$

with as many terms being retained as are necessary to describe the dependence. In this case, using only the first three terms, eqn 2.6 is rewritten as

$$\int dH = \int_{T_1}^{T_2} c_p dT = \int_{T_1}^{T_2} (a + bT + cT^2) dT$$

which, on integration, leads to

$$H_{T_2} - H_{T_1} = \{a(T_2 - T_1) + (b/2)(T_2^2 - T_1^2) + (c/3)(T_2^3 - T_1^3)\} \tag{2.12}$$

Mathematical note: Eqn 2.12 arises when the integration of each term is performed using the method outlined in Section 1.2.

For the majority of situations, the extra accuracy gained by this approach is barely significant. However, where large temperature ranges are involved, the additional effort in using eqn 2.12 is justified.

Example 2.5   Given the standard enthalpy of formation and heat capacity data below, calculate the enthalpy change at 100 °C for the gas phase reaction:

$$3\,C_2H_2 \rightarrow C_6H_6$$

$\Delta H_{f,298}^{\circ}/\text{kJ mol}^{-1} : C_2H_2\ 226.7;\ C_6H_6\ 82.9$

$$c_p(C_2H_2) = 30.7 + 5.28 \times 10^{-2}T - 1.63 \times 10^{-5}T^2 \text{ J mol}^{-1}\text{ K}^{-1}$$

$$c_p(C_6H_6) = -1.7 + 32.5 \times 10^{-2}T - 11.06 \times 10^{-5}T^2 \text{ J mol}^{-1}\text{ K}^{-1}$$

$$\Delta H_{298}^{\circ} = \Delta H_f^{\circ}(C_6H_6) - 3[\Delta H^{\circ}(C_2H_2)]$$
$$= 82.9 - 3(226.7) = -597.2 \text{ kJ mol}^{-1}$$

$$\Delta c_p = c_p(C_6H_6) - 3[c_p(C_2H_2)]$$
$$= (-1.7 + 32.5 \times 10^{-2}\ T - 11.06 \times 10^{-5}\ T^2)$$
$$- 3 \times (30.7 + 5.28 \times 10^{-2}\ T - 1.63 \times 10^{-5}\ T^2$$
$$= (-1.7 + 32.5 \times 10^{-2}\ T - 11.06 \times 10^{-5}\ T^2)$$
$$- (92.1 + 15.84 \times 10^{-2}\ T - 4.89 \times 10^{-5}\ T^2)$$
$$= (-1.7 - 92.1) + (32.5 \times 10^{-2}\ T - 15.84 \times 10^{-2}\ T)$$
$$- (11.06 \times 10^{-5}\ T^2 - 4.89 \times 10^{-5}\ T^2)$$
$$= -93.8 + 16.66 \times 10^{-2}\ T - 6.17 \times 10^{-5}\ T^2$$

Using the Kirchoff equation, eqn 2.5 we can therefore write

$$\Delta H_{373}^{\circ} = \Delta H_{298}^{\circ} + \int_{298}^{373} (-93.8 + 16.66 \times 10^{-2}\ T - 6.17 \times 10^{-5}\ T^2) dT$$

$$= -597.2 \times 10^3 \text{ J mol}^{-1} + \left[-93.8\,T + 16.66 \times 10^{-2} \frac{T^2}{2} - 6.17 \right.$$
$$\left. \times 10^{-5} \frac{T^3}{3} \right]_{298}^{373}$$

$$= -597.2 \times 10^3 \text{ J mol}^{-1} + -93.8(373 - 298)$$
$$+ 8.33 \times 10^{-2}(373^2 - 298^2) - 2.06 \times 10^{-5}(373^3 - 298^3)$$
$$= -597.2 \times 10^3 - 7035 + 4192 - 524 \text{ J mol}^{-1}$$
$$= -600.57 \text{ kJ mol}^{-1}$$

Exercise: The heat capacities at 298 K in $J\,K^{-1}\,mol^{-1}$ are 43.9 for ethyne and 81.7 for benzene. Use these values and eqn 2.8 to estimate ΔH° at 100 °C. Compare your answer with that from Example 2.5.

2.3 Enthalpy changes for other processes

Various other types of enthalpy change can be defined for a variety of processes which we encounter in chemistry.

Enthalpy of combustion. This is often referred to simply as the '*heat of combustion*' and is the enthalpy change when 1 mol of a compound reacts completely with excess oxygen gas. For example, the combustion of 1 mol of methane gas at 25 °C evolves 890.3 kJ.

$$CH_{4(g)} + 2O_{2(g)} \rightarrow CO_{2(g)} + 2H_2O_{(l)} \quad \Delta H_{298}^\circ(\text{combust.}) = -890.3 \text{ kJ mol}^{-1}$$

A variation of this that is commonly used is the *specific enthalpy*, J g^{-1}, which is the energy available from burning 1 g of a compound. It represents a measure of the usefulness of fuels, etc. which are often purchased by weight (or more correctly by mass!). For example, the specific enthalpy of carbon (in the form of graphite) is 32.8 kJ g^{-1}, that of saturated alkanes of the type that comprise the major portion of petrol is around 50 kJ g^{-1}. One of the highest values is that of hydrogen, 142 kJ g^{-1}, which explains its use as a rocket fuel. In mass terms, it would be one of the best sources of energy although other factors, primarily of economy and safety, mean that it is not widely used.

Enthalpies of combustion are relatively easy to measure as outlined in the next section. This leads to their common use to determine enthalpies of formation for compounds. As an example of the method, consider applying eqn 2.3 to the combustion of methane

Note: To a thermodynamicist, specific enthalpies are negative values since they involve the *evolution* of heat. However, this is usually taken for granted and the values are simply reported as the number, with the sign being understood.

$$CH_{4(g)} + 2O_{2(g)} \rightarrow CO_{2(g)} + 2H_2O_{(l)}$$
$$\Delta H^\circ(\text{combust.}) = [2\Delta H_f^\circ(H_2O) + \Delta H_f^\circ(CO_2)] - [\Delta H_f^\circ(CH_4) + 2\,\Delta H_f^\circ(O_2)]$$
$$-890.3 = [2(-285.8) + (-393.5)] - [\Delta H_f^\circ(CH_4) + 2(0)]$$
$$-890.3 = -965.1 - \Delta H_f^\circ(CH_4)$$
$$\Delta H_f^\circ(CH_4) = 74.8 \text{ kJ mol}^{-1}$$

Note: ΔH_f° (O_2) is zero since oxygen is an element in its standard state.

A good illustration of the use of ΔH°(combust.) to provide information indirectly on reactions that are difficult to study directly comes from biochemical systems (see Example 2.13). Our metabolic processes essentially consist of the reaction of food components with oxygen to form compounds such as water, carbon dioxide and urea, albeit conducted inside a living cell rather than a bomb calorimeter! Even at rest, a human body requires about

Note: Those of you on a diet may be worried that the figures here seem rather high compared with everyday experience. The problem is that the 'calories' reported by nutritionists and recorded on foodstuffs are in fact *kilo*calories.
1 'food calorie' = 4.18 *kilo*joules

6000 kJ per day which is provided by our food intake, and this is clearly much higher if you do anything other than lie in bed all day! As a guide, proteins and carbohydrates produce around 15–20 J g^{-1}, while the metabolism of fats yields 35–40 J g^{-1}.

Bond dissociation enthalpies.   These are often simply called *bond energies* and represent the energy per mole needed in the gas phase to break a chemical bond. For example, the dissociation of methane involves the breaking of 4 C–H bonds. Therefore

$$CH_{4(g)} \rightarrow C_{(g)} + 4H_{(g)} \quad \Delta H(\text{reaction}) = 4\Delta H_{(C-H)}$$

More correctly, $\Delta H_{(C-H)}$ should be called the *mean bond energy* since energy needed to break the C–H will depend on its precise environment in the molecule. For methane,

$$CH_3 - H \quad \Delta H_{(C-H)} = 432 \text{ kJ mol}^{-1}$$
$$CH_2 - H \quad \Delta H_{(C-H)} = 469 \text{ kJ mol}^{-1}$$
$$CH - H \quad \Delta H_{(C-H)} = 422 \text{ kJ mol}^{-1}$$
$$C - H \quad \Delta H_{(C-H)} = 339 \text{ kJ mol}^{-1} \quad \text{Average} = 1662/4 = 415.5 \text{ kJ mol}^{-1}$$

Table 2.1   Selected values of mean bond dissociation energies, $\Delta H_{(x-x)}$ in kJ mol^{-1}

C–C	348	C=C	615
C≡C	835	C–H	415
C–F	485	C–Cl	339
C–N	305	C–O	350
C=O	743	H–H	436
C$_{(graph)}$	717	F–F	155
Cl–Cl	242	Br–Br	193
Si–H	318	Si–Si	226
Si–O	466	O–O	146
O=O	498	O–H	464
N≡N	945	N–N	163

The final step in this sequence is a rather unusual reaction and therefore the value is rather out of line with the others. In the situations where we normally encounter C–H bonds, for example in organic compounds, the variation is relatively small and the bond energies can be used to estimate enthalpies of formation for compounds. However, care must always be taken to consider the type of bonding in the molecules and that the values shown in Table 2.1 can only be regarded as an approximation. These values are the average of values from a large number of compounds. Average bond energies can also be used to estimate enthalpy changes during reactions, as shown in Example 2.6.

In the case of a diatomic molecule, the bond energy is simply twice the enthalpy of formation of the isolated atoms

$$H_{2(g)} \rightarrow 2 \text{ H·} \quad \Delta H_{(H-H)} = 436 \text{ kJ mol}^{-1} = 2\Delta H_f^{\circ}(H\cdot)$$

Example 2.6   Use values of bond energies to estimate the standard enthalpy change for the reaction

$$CH_3OCH_{3(g)} \rightarrow C_2H_5OH_{(g)}$$

CH_3–O–CH_3 consists of 6 C–H bonds, and 2 C–O bonds. The energy required for the reaction

$$CH_3-O-CH_{3(g)} \rightarrow 6H_{(g)} + 2O_{(g)} + 2C_{(g)}$$

is therefore 6 $\Delta H_{(C-H)} + 2 \Delta H_{(C-O)} = 6(415) + 2(350)$
$$= 3190 \text{ kJ mol}^{-1}$$

The formation of CH_3CH_2OH requires 5 C–H bonds, 1 C–C, 1 C–O bonds and 1 O–H, so that the release of energy would be

$$5 \Delta H_{(C-H)} + \Delta H_{(C-C)} + \Delta H_{(C-O)} + \Delta H_{(O-H)} = 5(415) + (344) + (350)$$
$$+(465)$$
$$= -3234 \text{ kJ mol}^{-1}$$

The enthalpy change for the isomerization can therefore be estimated

$$\Delta H(\text{isomerization}) = \Delta H_1 + \Delta H_2 = 3190 + (-3234) = -44 \text{ kJ mol}^{-1}$$

$\Delta H^\circ_{f,298}$ for dimethyl ether and ethanol are -184.0 and -235.1 kJ mol^{-1}, respectively $\Delta H(\text{isomerization})$ is therefore -51.1 kJ mol^{-1}. While the agreement is not perfect, use of average bond energies typically gives a result with up to 10% error from the true value.

Enthalpies of phase change. The concept of phase equilibrium will be treated in detail in Chapter 5. For the moment, we will look at some changes of phase that will be familiar from everyday life.

The transition from solid to liquid, usually known as melting, is more correctly called *fusion*. Molecules in a liquid have much greater motion than those in a solid so that energy has to be added to the solid—fusion is an endothermic process. The enthalpy of fusion, ΔH^{fus} is the energy required at constant pressure to melt 1 mole of a pure component at its melting point, T_{m}. As usual, standard enthalpies of fusion are reported at 1 bar pressure. Some typical values are shown in Table 2.2.

An earlier name, still in common usage is that ΔH^{fus} is the *latent heat of fusion*. Since it essentially represents the energy needed to overcome the intermolecular forces in the solid, $\Delta H(\text{fusion})$ is often used as a measure of the strength of the bonding in solids. (Compare the values in Table 2.2 for sodium and silver.) It should be clear from previous sections of this chapter that, for the opposite process of solidification, or freezing,

$$\Delta H(\text{freezing}) = -\Delta H(\text{fusion}) \tag{2.13}$$

Table 2.2 Typical values of standard enthalpies of fusion and vaporization in kJ mol^{-1}. Melting, T_{m}, and boiling, T_{b}, points in K

	T_{m}	ΔH^{fus}	T_{b}	ΔH^{vap}
Helium	3.5	0.02	4.22	0.1
Argon	83.8	1.2	87.3	6.5
Methane	90.7	0.94	111.7	8.2
Ammonia	195.3	5.65	239.7	23.4
Ethanol	158.7	4.60	351.5	43.5
Water	273.2	6.01	373.2	40.7
Mercury	234.3	2.29	629.7	59.3
Sodium	371.0	2.60	1156	98.0
Silver	1234	11.3	2436	250.6

A second phase change commonly encountered is liquid $\rightarrow$ vapour, or *vaporization*. Following the usual pattern, ΔH^{vap} is the energy required at constant pressure to vaporize 1 mole of a pure liquid at its boiling point, T_{b}. Vaporization is therefore also endothermic. The values in Table 2.2 show that ΔH^{vap} is considerable larger than ΔH^{fus} since the molecules are completely separated in vaporization, whereas a considerable degree of intermolecular interaction remains in liquids. Vaporizing 1 mol of a liquid essentially involves overcoming all of the forces between the molecules, so ΔH^{vap} can be used as a measure of the strength of intermolecular forces in liquids. This is commonly expressed in a *cohesive energy density*, ced, which is the internal energy of vaporization per unit volume of liquid.

$$\text{ced} = \frac{\Delta U^{\text{vap}}}{V^\circ_{\text{m}}} = \frac{\Delta H^{\text{vap}} - RT}{V^\circ_{\text{m}}} \tag{2.14}$$

Some compounds do not display a liquid phase and the solid vaporizes directly to the gas, a process known as *sublimation*. An example is solid carbon dioxide, which, at atmospheric pressure, sublimates at -78 °C. This presents a straightforward application of Hess's law. At constant temperature,

$$\Delta H(\text{sublimation}) = \Delta H(\text{fusion}) + \Delta H(\text{vaporization}) \tag{2.15}$$

Example 2.7   Calculate the energy required to turn 100 g of ice at 0 °C into steam at 100 °C.

The process required can be conveniently split into three parts.

1.   100 g ice $\rightarrow$ 100 g water at 0 °C $\qquad \Delta H_1 = \Delta H^{\text{fus}}$ for 100 g
2.   100 g water, 0 °C $\rightarrow$ 100 g water, 100 °C $\quad \Delta H_2 = c_p \Delta T$ for 100 g
3.   100 g water, 100 °C $\rightarrow$ 100 g steam, 100 °C $\quad \Delta H_3 = \Delta H^{\text{vap}}$ for 100 g

$$100 \text{ g of water} = 100/18 \text{ mol} = 5.56 \text{ mol}$$

$$\begin{aligned} \Delta H_{\text{total}} &= \Delta H_1 + \Delta H_2 + \Delta H_3 \\ &= 5.56 \text{ mol}[+ 6.01 \times 10^3 + 75.1(373 - 273) + 40.7 \times 10^3] \\ &= 301.46 \text{ kJ } (100 \text{ g})^{-1} \end{aligned}$$

Note that the factor of 10^3 occurs in ΔH_1 and ΔH_3 since they are measured in kJ mol^{-1} while c_p is in J K^{-1} mol^{-1}.

Enthalpy of solution.   Many chemical reactions occur in solution and there are a number of types of enthalpy change associated with forming solutions. The most straightforward is *integral* enthalpy of solution, or the enthalpy change when 1 mol of a compound dissolves in a large excess of *pure solvent*.

$$\text{HCl}_{(g)} + (\text{aq}) \rightarrow \text{HCl}_{(aq)} \qquad \Delta H = 72.4 \text{ kJ mol}^{-1}$$

Here, the symbol (aq) signifies the presence of a large amount of water.

A related procedure is the dissolving of a compound in a large amount of *solution*. Here, the change is the *differential* enthalpy of solution and depends to a large extent on the concentration of the solution to which the compound is being added. For example, the dissolving of 1 mol of $\text{HCl}_{(g)}$ into 0.1 mol dm^{-3} hydrochloric acid solution will differ somewhat from that above, or for dissolving in a 1 mol dm^{-3} solution. However, in very dilute solutions, the value is quite constant.

Of particular importance, especially in biochemical systems, are processes occurring in aqueous solution including the reaction of ions with water—the *enthalpy of hydration* (see example 2.13).

2.4   Measurement of enthalpy changes

Enthalpy changes are measured directly in a *calorimeter*. This involves measuring the temperature change when known amounts of compounds react, usually in an insulated container. A complete listing of the various types of calorimeter would not be appropriate here but several types will be described to illustrate the basic principles.

In addition to direct calorimetry, other methods for measuring enthalpy changes in redox reactions are based on the variation of the electrochemical properties of the reaction. These will be discussed further in Chapter 4.

The bomb calorimeter.   The first example uses an indirect method for calculating *enthalpies of combustion*, ΔH(*combustion*). We saw in Section 1.4 that ΔU(combustion), i.e. the heat change at constant volume, could be conveniently measured in a bomb calorimeter. This value can be readily converted to ΔH(combustion) using the definition implicit in eqn 2.1

$$\Delta H(\text{combustion}) = \Delta U(\text{combustion}) + \Delta(pV)(\text{combustion}) \qquad (2.16)$$

Since the volume is constant, the last term in eqn 2.16 suggests that the pressure in the calorimeter must be measured. However, we can simplify the procedure by introducing the following assumptions.

Provided that the gases in the system behave according to the ideal gas equation, we can write

$$\Delta(pV) = (pV)_{\text{products}} - (pV)_{\text{reactants}} = (n_{\text{gas}}RT)_{\text{products}} - (n_{\text{gas}}RT)_{\text{reactants}}$$
$$\Delta(pV) = \Delta n_{\text{gas}}RT$$

Hence, $\qquad \Delta H(\text{combustion}) = \Delta U(\text{combustion}) + \Delta n_{\text{gas}}RT \qquad (2.17)$

where n_{gas} is the number of moles of gas involved in the reaction. The assumption of ideal gas behaviour is entirely reasonable for pressures up to several hundred bar pressures. The consideration only of gases in eqn 2.17 can be rationalized given the difference in the molar volumes between gases and compounds in their condensed phases.

To consider familiar values, the volume of 1 mole of an ideal gas at 25 °C and atmospheric pressure is 24 470 cm^3. The corresponding molar volumes for liquids and solids are much smaller. For example, at 25 °C water occupies $\sim$18 cm^3 mol^{-1}; hexane 132 cm^3 mol^{-1}; benzene 89 cm^3 mol^{-1}; and naphthalene $\sim$120 cm^3 mol^{-1}. Clearly then, neglect of these volumes in comparison with those of gases will cause only a small error in the ΔnRT term and will be negligible when used to convert ΔU to ΔH.

Example 2.8   Using the data from Section 1.4, calculate the enthalpy of combustion of naphthalene at 25 °C.

From Section 1.4, $\Delta U(\text{combustion})$ for naphthalene at 25 °C $= -5142$ kJ mol^{-1}

$$C_{10}H_{8(s)} + 12\ O_{2(g)} \rightarrow 10\ CO_{2(g)} + 4\ H_2O_{(l)}$$

To convert ΔH to ΔU using eqn 2.17, we need Δn_{gas}.

$$\Delta n_{\text{gas}} = n_{\text{gas}}(\text{products}) - n_{\text{gas}}(\text{reactants}) = 10 - 12 = -2\ \text{mol}$$

Hence,

$$\Delta H(\text{combustion}) = -5\,142\,000\ \text{J mol}^{-1} + (-2\ \text{mol})$$
$$(8.314\ \text{J K}^{-1}\text{mol}^{-1})(298.2\ \text{K})$$
$$\Delta H(\text{combustion}) = -5\,147\,000\ \text{J mol}^{-1} = -5147\ \text{kJ mol}^{-1}$$

Note that the difference between the internal energy and enthalpy is relatively small in this case.

Exercise: Take the values from Example 2.8 and calculate the error in neglecting the volumes of naphthalene and water.

Flame calorimeter.   The above apparatus is not suitable for measuring the properties of gas reactions. These are usually performed in a flame calorimeter, illustrated schematically in Fig. 2.4.

The gases concerned are fed at a known, constant rate to a jet at which the reaction occurs. The reaction chamber and gas pipes are contained in a thermostatically controlled water bath to ensure constant temperature. The

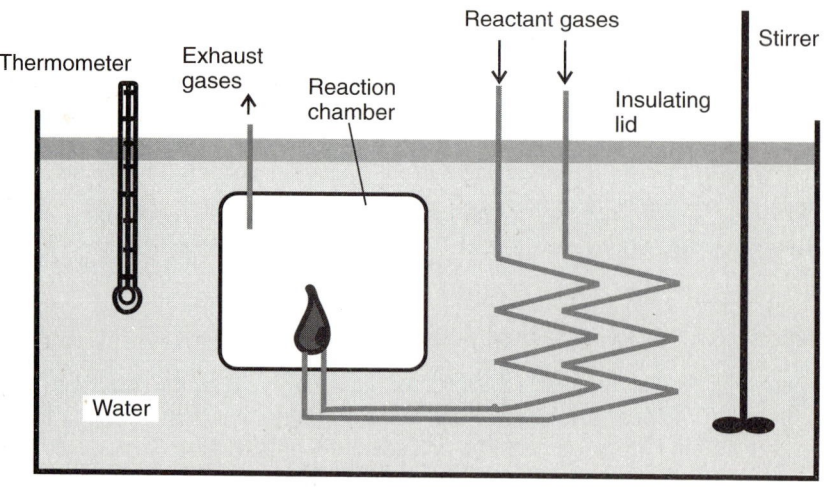

Fig. 2.4   Schematic diagram of a flame calorimeter.

reaction is then started and the temperature rise measured after a known amount of gas has been fed into the reaction. Calibration of the calorimeter, either with a standard reaction or by electrical means (similar to that described in Section 1.4 for the bomb calorimeter), allows calculation of ΔH(reaction) since the reaction is conducted at constant pressure. Example 2.9 illustrates the use of this type of apparatus for measuring an enthalpy of combustion where one of the gases used is oxygen.

The chemical structures of *n*-butane and isobutane (2-methyl propane)

CH₃—CH₂—CH₂—CH₃   (n - butane)

(CH₃)₂CH—CH₃   (isobutane)

n - butane   **isobutane**

Example 2.9   The combustion of 2.9 g of *n*-butane gas, C_4H_{10}, raised the temperature of a flame calorimeter by 11.47 °C. The enthalpy of combustion at 25 °C of n-butane is -2877.1 kJ mol^{-1}.

The gas flow was then changed to isobutane. Combustion of a total volume of 448 cm^3 (measured at STP) at 25 °C caused a temperature rise of 4.575 °C. Calculate the enthalpy of combustion of isobutane.

The relative molar mass of *n*-butane is 58 g mol^{-1}. Hence 2.9 g is 0.05 mol, so that the energy released on combustion is $(0.05 \text{ mol})(-2877.1 \text{ kJ mol}^{-1})$ $= -143.86$ kJ.

Thus, a temperature rise of 11.47 °C was caused by -143.86 kJ.

1 °C represents an energy release of -12.54 kJ.

The 4.575 °C generated by the isobutane is equivalent to an energy release of -57.37 kJ, which occurs from 448 cm^3 or $(448/22\ 400) = 0.02$ mol

Enthalpy of combustion $= -57.37$ kJ/0.02 mol $= -2868.5$ kJ mol^{-1}

Note that the difference in the two enthalpies of combustion is equivalent to the enthalpy of isomerization for the two butane isomers.

Solution calorimeter. This type of apparatus is commonly used for measuring heat changes during reactions that occur in solution and is illustrated in Fig. 2.5. It allows two (or more) reactants to be placed in an insulated, thermostatted chamber until thermal equilibrium is reached.

The reactants can then be rapidly mixed to start the reaction, either by tilting the reaction vessel, or, in more sophisticated versions, using a magnet to break a glass seal between the chambers. Calibration of this type of apparatus is often carried out using a neutralization reaction between a strong acid and a strong base. The enthalpy change for this reaction is -56.9 kJ mol^{-1}.

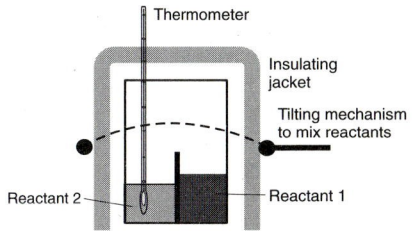

Fig. 2.5 Schematic diagram of a calorimeter suitable for measuring enthalpy changes for reactions in solution. An alternative design uses an external magnet to remove a partition between the two reactants.

Example 2.10 The reaction between 50 cm^3 of 0.10 mol dm^3 hydrochloric acid solution and the same volume of 0.10 mol dm^3 sodium hydroxide solution in a solution calorimeter caused a temperature increase of 0.795 °C.

100 cm^3 of water and 2.500 g of anhydrous sodium carbonate were then mixed in the calorimeter, resulting in a temperature increase of 1.55 °C. The experiment was repeated using 3.50 g of sodium carbonate decahydrate and the temperature decreased by 2.310 °C.

Calculate the enthalpy of solution for both compounds and the enthalpy of hydration of sodium carbonate.

---

50 cm^3 of 0.10 mol dm^3 HCl solution represents 5×10^{-3} moles.

The enthalpy change is -56.9 kJ mol^{-1} so the energy released is 284.5 J. This raised the temperature by 0.795 °C. Hence, the 'heat capacity' of the calorimeter is 284.5 J/0.795 °C = 357.9 J K^{-1}.

The reaction of Na$_2$CO$_{3(s)}$ caused a temperature rise of 1.55 °C. Hence, from the calibration, this represents an energy release of (1.55 K $\times$ 357.9 J K^{-1}) = 554.75 J. 2.50 g = (2.5/105.99) = 0.0236 mol.

∴ enthalpy of solution = (-554.75 J/0.0236 mol) = $-23\ 506$ J mol^{-1}.

Considering Na$_2$CO$_3$(s)·10 H$_2$O, 3.50 g is 0.01223 mol. The energy change is (2.31 K $\times$ 357.9 J K^{-1}) = 826.75 J. In this case, the temperature *decreases* so that the enthalpy change is positive.

∴ enthalpy of solution = ($+826.75$ J/0.01223 mol) = $67\ 600$ J mol^{-1}.

We can write the two reactions as:

$$Na_2CO_{3(s)} + \infty H_2O_{(l)} \rightarrow Na_2CO_{3(aq)} \quad \Delta H = -23.506 \text{ kJ mol}^{-1}$$
$$Na_2CO_3 \cdot 10H_2O_{(s)} + \infty H_2O_{(l)} \rightarrow Na_2CO_{3(aq)} \quad \Delta H = +67.6 \text{ kJ mol}^{-1}$$

If we subtract these two equations, we get

$$Na_2CO_{3(s)} \rightarrow Na_2CO_3 \cdot 10H_2O_{(s)} \quad \Delta H = (-23.506) - (+67.6) \text{ kJ mol}^{-1}.$$

$$\text{Enthalpy of hydration} = -91.106 \text{ kJ mol}^{-1}$$

Note that, strictly, this example only works since the same volume of solution was used for all the experiments. If this is not the case, allowance must be made for the change in heat capacity.

Note: The reason that the calibration reaction works irrespective of the acid and base used is that all neutralizations involve the same reaction, which can effectively be represented as

$$H^+_{(aq)} + OH^-_{(aq)} \rightarrow H_2O_{(l)}$$

The other ions do not take part in the reaction and so are unimportant.

2.5 Concluding remarks

This chapter has introduced the basic ideas of thermochemistry. There are many instances where knowledge of the heat changes during reactions is important. The examples presented have illustrated particular aspects of the ideas introduced but we conclude this chapter with four further examples which show how the various ideas and concepts can be fitted together in real chemical situations.

The first of these (Example 2.11) illustrates how the rate of heat evolution from a reaction can be calculated under a non-standard set of conditions. This is a simplified form of an important calculation which must be made when considering the design and safe operation of chemical plants. Example 2.12 gives further applications of mean bond energies and again illustrates the care which must be taken in their use. Example 2.13 shows how the enthalpy changes for sequences of reactions can be combined to give information on reactions which cannot be readily carried out, in this case a reaction being one of biochemical interest. The final example (Example 2.14) further illustrates the use of a bomb calorimeter to measure the calorific value of a food, which has also been suggested as a future fuel stock.

Note: The first stage of the solution is to calculate the enthalpy change at 25 °C. This can then be converted to 450 °C using the mean heat capacities. The rate of reaction then leads to the rate of heat evolution.

Example 2.11   A gas mixture consisting of 25% nitrogen and 75% hydrogen (by volume) was passed over a catalyst at a rate of 112 dm^3 min^{-1} (gas volumes corrected to 0 °C and 1 atm pressure). The reaction took place at 450 °C and 1 bar pressure and complete conversion to ammonia was achieved. Given the following data, calculate the rate of heat evolution or absorption.

$\Delta H^\circ_{f,298}(NH_3)$ − 46.0 kJ mol^{-1}

Mean $\bar{c}_p$/J K^{-1} mol^{-1} :   N$_2$ 29.7 : H$_2$ 29.3; NH$_3$ 39.7

$$N_{2(g)} + 3H_{2(g)} \rightarrow 2NH_{3(g)}$$

The enthalpy change at 25 °C can be calculated from eqn 2.3.

$$\Delta H^\circ_{298} = [2 \times \Delta H^\circ_{f,298}(NH_3)] - [\Delta H^\circ_{f,298}(N_2) + 3\Delta H^\circ_{f,298}(H_2)]$$
$$= [-92.0] - [0] \text{ kJ mol}^{-1}$$
$$\Delta H^\circ_{298} = -92.0 \text{ kJ mol}^{-1}$$

Correcting the enthalpy change to 450 °C makes use of eqn 2.8.

$$\Delta H^\circ_T(\text{reaction}) = \Delta H^\circ_{298}(\text{reaction}) + \Delta\bar{c}_p(T - 298.15)$$
$$\Delta\bar{c}_p = [2 \times \bar{c}_p(NH_3)] - \bar{c}_p(N_2) + 3 \times \bar{c}_p(H_2)]$$
$$= (2 \times 39.7) - [29.7 + (3 \times 29.3)]$$
$$= -38.2 \text{ J K}^{-1} \text{ mol}^{-1}$$

$$\Delta H^\circ_{723} = 92\,000 \text{ J mol}^{-1} + (-38.2 \text{ J K}^{-1} \text{ mol}^{-1})(723.15 - 298.15)\text{K}$$
$$= -92\,000 \text{ J mol}^{-1} + (-16235 \text{ J mol}^{-1})$$
$$\Delta H^\circ_{723} = -108.24 \text{ kJ mol}^{-1}$$

Hence, at 450 °C, 108.24 kJ is evolved for each mole of the equation as written. From the stoichiometry, this means that 108.24 kJ is evolved per mole of nitrogen that reacts. Since, for a gas, the volume is proportional to the number of moles, the 1 : 3 mixture is in the stoichiometric amount. We know that at 0 °C and 1 atm pressure, 1 mole of an ideal gas occupies 22.4 dm^3. Thus, the total amount of gas passing over the catalyst amounted to 5 moles per minute, of which 1.25 moles were N_2.

Rate of heat evolution = (moles of N_2 min^{-1}) (heat evolved per mole N_2)

$$= (1.25 \text{ mol min}^{-1}) (-108.24 \text{ kJ (mol } N_2^{-1})$$
$$= -135.3 \text{ kJ min}^{-1}$$

The negative sign indicates an exothermic reaction so that heat is evolved.

Example 2.12 Using the data in Table 2.1, estimate the standard enthalpies of formation for (a) gaseous cyclohexane, C_6H_{12}, and (b) gaseous benzene, C_6H_6. Compare the answers with the values in Appendix 1.

Benzene

Cyclohexane

The chemical structures of benzene and cyclohexane.

(a) The formation of cyclohexane can be written

$$6C_{(graph)} + 6H_{2(g)} \rightarrow C_6H_{12(g)} \quad \Delta H_1^\circ = \Delta H_f^\circ(C_6H_{12(g)})$$

To form cyclohexane, we need to 'make' 6 C–C bonds and 12 C–H bonds. From the definition of bond energies, this refers to the reaction

$$6C_{(g)} + 12H_{(g)} \rightarrow C_6H_{12(g)} \quad \Delta H_2^\circ = 6(-344) + 12(-415)$$
$$= -7044 \text{ kJ mol}^{-1}$$

the values being negative since this is the reverse of bond dissociation.

Prior to this step, we need

$$6C_{(graph)} + 6H_{2(g)} \rightarrow 6C_{(g)} + 12H_{(g)} \quad \Delta H_3^\circ = 6\Delta H_{(graph)} + 6\Delta H_{(H-H)}$$
$$\Delta H_3^\circ = 6(716.7) + 6(436) = 6916.2 \text{ kJ mol}^{-1}.$$

It is clear that $\Delta H_1^\circ = \Delta H_2^\circ + \Delta H_3^\circ = -7030 + 6916.2 = -127.8$ kJ mol^{-1}

The actual value is -123.2 kJ mol^{-1}. The agreement is therefore very good considering that average bond energies have been used throughout.

(b) This calculation is performed in an analogous manner to part (a) except that benzene contains 6 C–H, 3 C–C, and 3 C=C bonds.

$$6C_{(graph)} + 3H_{2(g)} \rightarrow C_6H_{6(g)} \quad \Delta H_1^\circ = \Delta H_f^\circ(C_6H_6(g))$$
$$6C_{(g)} + 6H_{(g)} \rightarrow C_6H_{6(g)} \quad \Delta H_2^\circ = 3(-615) + 3(-344) + 6(-415)$$
$$= -5367 \text{ kJ mol}^{-1}$$
$$6C_{(graph)} + 3H_{2(g)} \rightarrow 6C_{(g)} + 6H_{(g)} \quad \Delta H_3^\circ = 6\Delta H_{(graph.)} + 6\Delta H_{(H-H)}$$
$$\Delta H_3^\circ = 6(716.7) + 3(436) = 5608.2 \text{ kJ mol}^{-1}.$$

It is clear that $\Delta H_1^\circ = \Delta H_2^\circ + \Delta H_3^\circ = -7030 + 6916.2 = 241.2 \text{ kJ mol}^{-1}$
   The actual value is 83 kJ mol^{-1}. Clearly, there is poor agreement for benzene. This is due to the assumption of the alternating single and double bonds in benzene whereas the 'actual' structure is a resonance-stabilized hybrid of these.

Note that this illustrates that care must be taken to consider the chemical properties of the molecules involved when using average bond enthalpy data. At best, approximate results can be expected.

Example 2.13   The major products from the metabolism of amino acids are urea, carbon dioxide, and water. The enthalpy change for the reaction of glycine to $CO_{2(g)}$, water, and ammonia gas at 25 °C is -1163.2 kJ mol^{-1}. The enthalpy change on hydrolysis of solid urea to $CO_{2(g)}$, water, and ammonia is 133.6 kJ mol^{-1}. The enthalpies of solution for glycine and urea are 15.69 and 13.93 kJ mol^{-1}, respectively.
   Calculate the enthalpy change for the reaction of glycine with oxygen to form urea in aqueous solution.

The reaction which is the target of the question is

$$2NH_2CH_2COOH_{(aq)} + 3O_{2(g)} \rightarrow NH_2CONH_{2(aq)} + 3H_2O_{(l)} + 3CO_{2(g)}$$

The data given can be represented by:

(a) $2NH_2CH_2COOH_{(s)} + 3O_{2(g)} \rightarrow 2NH_{3(g)} + 2H_2O_{(l)} + 4CO_{2(g)}$
$$\Delta H = -1172.0 \text{ kJ mol}^{-1}$$

(b) $NH_2CONH_{2(s)} + H_2O_{(l)} \rightarrow 2NH_{3(g)} + CO_{2(g)}$
$$\Delta H = 133.6 \text{ kJ mol}^{-1}$$

(c) $NH_2CH_2COOH_{(s)} + \infty H_2O_{(l)} \rightarrow NH_2CH_2COOH_{(aq)}$
$$\Delta H = -15.69 \text{ kJ mol}^{-1}$$

(d) $NH_2CONH_{2(s)} + \infty H_2O_{(l)} \rightarrow NH_2CONH_{2(aq)}$
$$\Delta H = 13.93 \text{ kJ mol}^{-1}$$

If we subtract (b) from (a), we get

$$[2NH_2CH_2COOH_{(s)} + 3O_{2(g)}] - [NH_2CONH_{2(s)} + H_2O_{(l)}]$$
$$\rightarrow [2NH_{3(g)} + 2H_2O_{(l)} + 4CO_{2(g)}] - [2NH_{3(g)} + CO_{2(s)}]$$

or

(e) $2NH_2CH_2COOH_{(s)} + 3O_{2(g)} \rightarrow NH_2CONH_{2(s)} + 3H_2O_{(l)} + 3CO_{2(g)}$
$$\Delta H = (-1172) - (133.6) = -1305.6 \text{ kJ mol}^{-1}$$

This is for the reaction as written with solid glycine and urea. To convert these values to those in solution, we have to make use of (c) and (d).

If we take (e) − {2 × (c)} + (d),

$$[2NH_2CH_2COOH_{(s)} + 3O_{2(g)}] - [2NH_2CH_2COOOH_{(s)} + \infty H_2O_{(l)}] + $$
$$[NH_2CONH_{2(s)} + \infty H_2O_{(l)}] \rightarrow$$
$$[NH_2CONH_{2(s)} + 3H_2O_{(l)} + 3CO_{2(g)}] - [2NH_2CH_2COOOH_{(aq)}] + $$
$$[NH_2CONH_{2(aq)}]$$

which reduces to

$$3O_{2(g)} + NH_2CONH_{2(s)} \rightarrow NH_2CONH_{2(s)} + 3H_2O_{(l)} + 3CO_{2(g)}$$
$$-2NH_2CH_2COOH_{(aq)} + NH_2CONH_{2(aq)}$$

or

$$2NH_2CH_2COOH_{(aq)} + 3O_{2(g)} \rightarrow 3H_2O_{(l)} + 3CO_{2(g)} + NH_2CONH_{2(aq)}$$

which is the target. ΔH is therefore given by $\{\Delta H_{(e)} - (2 \times \Delta H_{(c)}) + \Delta H_{(d)}\}$

$$\Delta H = (-1305.6) - (2 \times 15.69) + (13.93)\ \text{kJ mol}^{-1}$$
$$= -1323.05\ \text{kJ mol}^{-1}$$

In this case, the difference between the reactions for the solid reagents and those in solution is not very great. However, this is not always the case.

Example 2.14 The main form of sugar we use as a food is sucrose, $C_{12}H_{22}O_{11}$. The complete combustion of 2.0026 g of sucrose in a bomb calorimeter with a heat capacity of 11 140 JK^{-1} at 25 °C resulted in a temperature increase of 2.966 °C. Calculate $\Delta H^{\text{combustion}}$ for sucrose.

The energy released on combustion is given by $(11\ 140\ \text{J K}^{-1} \times 2.966\ \text{K})$ = 33.041 kJ

Since the bomb calorimeter is a constant-volume device, this gives $\Delta U^{\text{combustion}}$. The experiment used 5.85×10^{-3} moles.

Hence,

$$\Delta U^{\text{combustion}} = -33.041/5.85 \times 10^{-3} = -5647.5\ \text{kJ mol}^{-1}$$

The chemical equation for the reaction is

$$C_{12}H_{22}O_{11(s)} + 12O_{2(g)} \rightarrow 12CO_{2(g)} + 11H_2O_{(l)}$$

so the number of moles of gas is the same before and after the reaction. Hence, $\Delta n_{\text{gas}} = 0$ and $\Delta H^{\text{combustion}} = \Delta U^{\text{combustion}}$

$$\therefore \Delta H^{\text{combustion}} = -5647.5\ \text{kJ mol}^{-1}$$

3 Entropy in chemistry

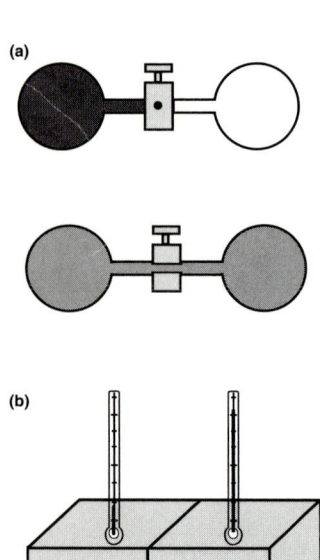

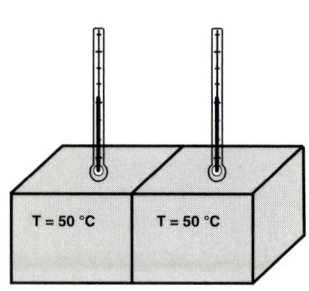

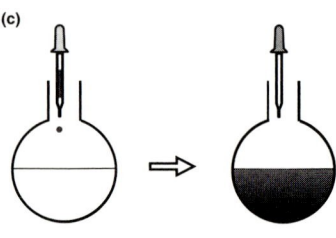

Fig. 3.1 Spontaneous processes. See text for explanation.

In general, in chemistry we are concerned with the factors that determine whether reactions or processes can occur, and, if so, how far and how fast they proceed. Thermodynamics can tell us nothing about the rate of reactions but can tell us everything about the other factors.

In the early days of physical chemistry it was thought that systems reacted or underwent change so as to minimize their energy. However, things are not quite that simple. Consider, for example the dissolving of ammonium nitrate in water, which is endothermic. Clearly, here the system gains energy in absorbing heat but the reaction occurs readily. Also, consider the expansion of an ideal gas at constant temperature. There is no change in internal energy but it happens anyway. There are a number of other examples which we encounter in everyday life of things that we know will happen in one direction only, some of which are illustrated in Fig. 3.1. Such processes are *spontaneous*.

If we open the tap in Fig. 3.1a, then gas will flow into the evacuated vessel. It is very unlikely that, at some future time all the gas will rediffuse into the left-hand bulb and leave a vacuum in the right-hand bulb. If we take a hot metal block and place it against a cold one, as in Fig. 3.1 (b), we know that the temperatures will equalize; it is very unlikely that at a later time one end will spontaneously get hotter than the other. If we place a drop of ink into a bath of water, it will gradually spread through the bath to give a constant concentration. It is then unlikely that all the ink will diffuse into one place in the bath. As a final chemical example, we know that hydrogen reacts very well with oxygen to form water. However, around room temperature the reverse reaction does not take place spontaneously.

Clearly then, there is some factor other than the energy which plays a part in determining whether reactions happen, and this is provided by the *entropy*, given the symbol S. There are a number of ways of looking at this, but the most relevant to the present discussion is to consider the molecular state of the system.

In each of the physical situations described above (we will return to the water reaction later) the final situation results in the gas molecules, the heat, or the ink molecules being distributed over a greater amount of space, leading to a more random distribution of matter. We describe the entropy as a measure of this randomness or *disorder* of the system, which clearly increases in the processes shown in Fig. 3.1. A fundamental definition of entropy is given by the *Boltzmann equation*.

$$S = k_B \ln \omega \tag{3.1}$$

where ω is the number of ways of arranging the molecules of the system and k_B is a constant, known as the *Boltzmann constant*. Thus, the larger the number of arrangements, or the less organized the system, the larger the entropy. This equation forms the basis of the powerful methods of *statistical thermodynamics*.

3.1 The Second Law of thermodynamics

The main usefulness of entropy in chemistry comes from the ability to use the concept to predict the direction of chemical (or any other) change. This arises as a result of a large number of empirical observations such as those described above and is embodied in the *Second Law of thermodynamics*. This can be stated in a number of ways, but perhaps the most straightforward is:

Spontaneous processes are those which INCREASE *the entropy of the universe.*

This may seem a rather grand definition but we will see shortly that it is not necessary to consider the complete universe every time we think about a reaction. The key word here is *spontaneous*. If we can determine the change of entropy for a process, we will be able to predict its direction and the remainder of this chapter is devoted to these methods. From the idea of entropy being related to order, we can conclude that spontaneous processes are those which lead to an increase in disorder.

3.2 A quantitative measure of entropy

Historically, the development in the 19th century of quantitative models of the entropy of systems relied on the consideration of heat–work cycles and the efficiency of heat engines. However, this has limited usefulness and gives limited insight into chemical situations and we will simply state the result and justify it with some examples.

$$\Delta S = \frac{q_{rev}}{T} \qquad (3.2)$$

where ΔS is the entropy *change* when an amount of heat q_{rev} is added in a reversible manner at temperature T. As noted in Section 1.2, q_{rev} is the maximum heat change available so that the entropy is associated with the maximum energy change which a system can undergo.

Although not formally proved, eqn 3.2 is entirely consistent with everything that has been said so far about entropy. Heating a system leads to faster motion of molecules, increased bond vibration, and larger population of higher energy levels. This leads to the system having greater disorder. Adding the same amount of heat would be expected to have a greater effect at lower temperatures, hence the inverse temperature dependence. Finally, the heating must be carried out so that the temperature of the system is uniform throughout, hence the requirement for a reversible change.

Having introduced eqn 3.2, we can now apply it to a number of situations of chemical importance. Although it may not be obvious from the preceding discussion, it can be shown that entropy is a state function so that again we can use the form

$$\Delta S = S_{final} - S_{initial} \qquad (3.3)$$

without worrying about the path by which the change is carried out.

Entropy changes during a change of phase

This provides a straightforward application of eqn 3.2 since, at the temperature of the phase change, the heat change can be considered reversible at constant pressure. If we take vaporization and fusion, then

$$\Delta S^{\text{vap}} = \frac{\Delta H^{\text{vap}}}{T_{\text{b}}} \quad ; \quad \Delta S^{\text{fus}} = \frac{\Delta H^{\text{fus}}}{T_{\text{m}}} \tag{3.4}$$

Note: Eqn 3.4 will be derived again in Chapter 5 starting from a different point of view.

where ΔH^{vap} and ΔH^{fus} are the enthalpies of vaporization and fusion at the boiling and melting temperatures, T_{b} and T_{m}, respectively. Clearly, ΔS for the reverse changes—condensation or freezing—are given by the negative of the values calculated according to eqn 3.4.

As discussed in Section 2.3, ΔH^{vap} and ΔH^{fus} are endothermic. Thus, ΔS must be positive for these phase changes. This is consistent with our idea of entropy as a measure of disorder. A vapour must be more disordered and so have higher entropy than a liquid, so ΔS^{vap} is positive. The same argument holds for melting. Further discussion of these changes will be undertaken in Chapter 5.

Isothermal expansion of an ideal gas

While this is perhaps the least important from the point of view of a chemical reaction, it provides another straightforward example of applying eqn 3.2.

It was shown in Section 1.2 that, for an isothermal change in an ideal gas, $\Delta U = 0$ so $q_{\text{rev}} = -w_{\text{rev}}$. Also, from eqn 1.5, $w_{\text{rev}} = -nRT \ln (V_{\text{final}}/V_{\text{initial}})$.

$$\text{Hence,} \quad \Delta S = S_{V_{\text{final}}} - S_{V_{\text{initial}}} = \frac{q_{\text{rev}}}{T} = \frac{-w_{\text{rev}}}{T} = nR \ln \left(\frac{V_{\text{final}}}{V_{\text{initial}}} \right) \tag{3.5}$$

If a gas expands, $V_{\text{final}} > V_{\text{initial}}$ so the entropy change is positive. This is again consistent with our idea of entropy as a measure of disorder since there is now a greater volume over which to arrange the gas molecules. Given the inverse relation between pressure and volume, eqn 3.5 could be written as

$$\Delta S = S_{p_{\text{final}}} - S_{p_{\text{initial}}} = nR \ln \left(\frac{p_{\text{initial}}}{p_{\text{final}}} \right)$$

Effect of temperature on entropy

Again, we make use of a relation used previously; $q_{\text{rev}} = c_{\text{p}} \, dT$ at constant pressure. The entropy at some temperature T_{f} can be related to the entropy at a different temperature, T_{i} by

$$\Delta S = S_{T_{\text{f}}} - S_{T_{\text{i}}} = \frac{dq_{\text{rev}}}{T} = \int_{T_{\text{i}}}^{T_{\text{f}}} \frac{\bar{c}_{\text{p}}}{T} dT \tag{3.6}$$

If, as in our discussion of enthalpy changes, we begin by approximating that $\bar{c}_{\text{p}}$ is constant over the temperature range involved, then providing no change of phase occurs:

$$S_{T_{\text{f}}} = S_{T_{\text{i}}} + \bar{c}_{\text{p}} \int_{T_{\text{i}}}^{T_{\text{f}}} \frac{1}{T} dT = \bar{c}_{\text{p}} \ln \left(\frac{T_{\text{f}}}{T_{\text{i}}} \right) \tag{3.7}$$

Equation 3.7 is adequate over relatively small temperature ranges where $\bar{c}_{\text{p}}$ is the mean value over the range. If there is a significant variation in c_{p} then we must proceed in an analogous manner to that of Section 2.2. Equation 3.6 can be rewritten to include the power series representation of c_{p}.

$$S_{T_f} = S_{T_i} + \int_{T_i}^{T_f} \frac{c_p(T)}{T}\,dT = S_{T_i} + \int_{T_i}^{T_f} \frac{a + bT + cT^2}{T}\,dT$$

$$S_{T_f} = S_{T_i} + \left[a \ln T + bT + (c/2)T^2\right]_{T_i}^{T_f} = a \ln\left(\frac{T_f}{T_i}\right) + b(T_f - T_i) + (c/2)\left(T_f^2 - T_i^2\right)$$

Experimentally, the most straightforward method for applying eqn 3.6 is to use a graphical technique. If values of c_p are measured as a function of temperature, a graph of (c_p/T) versus temperature can be constructed. Finding the increase of entropy then amounts to finding the area under the curve, as shown in Fig. 3.2.

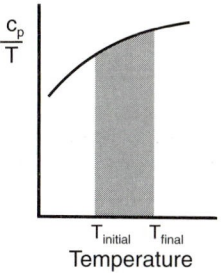

Fig. 3.2 Graphical determination of the entropy as a function of temperature. ΔS is given by the shaded area under the curve between the initial and final temperatures.

Example 3.1 Given that the heat capacities of ice and water at 0 °C are 2.2 and 4.18 J K^{-1} g^{-1} and the enthalpy of fusion is 332 J g^{-1}, calculate the entropy change for the freezing of 1 mole of supercooled water at -10 °C.

Clearly, $H_2O_{(l)}$, $-10°C \to H_2O_{(s)}$, $-10°C$ cannot be carried out in a reversible manner. Thus, in order to use eqn 3.2 to calculate ΔS, we must separate the process into a series of reversible changes and sum ΔS for each. This procedure is valid since entropy is a state function.

$$\text{water, } -10\,°C \xrightarrow{\Delta S_1} \text{water, } 0\,°C \xrightarrow{\Delta S_2} \text{ice, } 0\,°C \xrightarrow{\Delta S_3} \text{ice, } -10\,°C$$

$$\Delta S = \Delta S_1 + \Delta S_2 + \Delta S_3 = 4.18 \ln\left(\frac{273}{263}\right) + \frac{-332}{273} + 2.2 \ln\left(\frac{263}{273}\right) \text{ J K}^{-1}\text{g}^{-1}$$

$$\Delta S = 0.156 + (-1.216) + (0.082) = -1.142 \text{ J K}^{-1}\text{g}^{-1}$$

Hence, $\Delta S = (18 \text{ g mol}^{-1})(-1.142 \text{ J K}^{-1}\text{ g}^{-1}) = -20.556 \text{ J K}^{-1}\text{ mol}^{-1}$

Note that we have a negative entropy change arising from a liquid forming a more ordered solid.

3.3 Absolute entropies and the Third Law

From eqns 3.4 and 3.6, we can now calculate the change in entropy for any compound when it undergoes a change of phase or temperature. However, as yet we cannot define the entropy of a compound in absolute terms. In order to do this, as in our discussion of energy, we need a standard to which we can refer values. In this case though, different elements will have different degrees of order at a given temperature. Consider oxygen, mercury, and copper—can they be said to have the same degree of atomic and molecular order at 25 °C? Thus, we cannot use the same standard as when defining $\Delta H^\circ_{f,298}$.

To give a standard to act as a zero for entropy, we need a perfectly ordered system. Although it can never be achieved in practice, it is clear that this condition would be fulfilled by a perfect solid crystal at absolute zero temperature so that there is no motion of any type. A perfect crystal is one with all the molecules aligned perfectly and no defects. We can therefore postulate the *Third Law of thermodynamics* as:

The entropy of a perfect crystal at zero Kelvin is zero

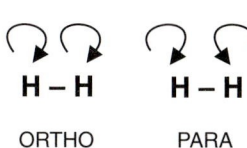

ORTHO   PARA

Note: The two forms of hydrogen have different *nuclear* spins on the protons. (This should not be confused with *electron* spins.) The arrow indicates the direction of the spins.

```
C–O C–O C–O      C–O C–O C–O
C–O C–O C–O      C–O C–O C–O
C–O C–O C–O  ≈  C–O C–O C–O
C–O C–O C–O      C–O C–O C–O
C–O C–O C–O      C–O C–O C–O
```

Fig. 3.3 Different orientations in a solid crystal of carbon monoxide.

In practice, a perfect crystal can never be achieved. All crystals contain defects. Even when all the molecules are aligned perfectly, other more subtle effects can result in a *residual entropy* at 0 K. For example, molecules containing different isotopes (e.g. $H^{35}Cl$ and $H^{37}Cl$) may be present. In hydrogen, the mixture of *ortho* and *para* forms (which have different pairing of nuclear spins) means that $S \neq 0$. The dipole in CO is very small so that there is a negligible difference in energy in the orientation of molecules in the crystal. Thus, even at the lowest temperatures, the solid consists of an array of molecules with random orientation (Fig. 3.3).

The residual entropy caused by this effect can be estimated from eqn 3.1. If both orientations are equally likely (i.e. there is zero energy difference) then for N molecules there are 2^N possible ways of arranging them. From the Boltzmann equation and putting N = Avogadro's number for 1 mole of CO,

Mathematical note: $\ln x^a = a \ln x$

$$S = k_B \ln \omega = (1.38 \times 10^{-23} \text{ J}) \ln 2^N$$

$$= (1.38 \times 10^{-23} \text{ J K}^{-1})(6.023 \times 10^{23} \text{ mol}^{-1}) \ln 2$$

$$= 5.76 \text{ J K}^{-1} \text{ mol}^{-1}$$

In fact, precise spectroscopic measurements show that the residual entropy of CO is 4.6 J K^{-1} mol^{-1} indicating that the arrangement is not, in fact, completely random.

Standard or absolute entropies

For convenience, tables of entropy values are published under standard conditions. Some example values are given in Table 3.1 and a more comprehensive list in Appendix 1.

The standard entropy, S_{298}° of a compound is the molar entropy at 298.15 K and 1 bar pressure

Table 3.1 Typical values of S_{298}°/J K^{-1} mol^{-1}

$C_{(s)}$ (graphite)	5.7
$C_{(s)}$ (diamond	2.4
$Cu_{(s)}$	33.3
$NaCl_{(s)}$	72.4
Glycine, $NH_2CH_2CO_2H_{(s)}$	103.5
D-Glucose, $C_6H_{12}O_{6(s)}$	212.0
Sucrose, $C_{12}H_{22}O_{11(s)}$	360.2
$Hg_{(l)}$	77.4
Benzene, $C_6H_{6(l)}$	124.3
Ethanol, $C_2H_5OH_{(l)}$	160.7
$H_{2(g)}$	130.7
$CO_{(g)}$	197.3
$O_{2(g)}$	205.1
$CO_{2(g)}$	213.6
Benzene, $C_6H_{6(g)}$	269.2
Ice, $H_2O_{(s)}$	45.0
Water, $H_2O_{(l)}$	69.9
Water vapour, $H_2O_{(g)}$	188.8

Unlike internal energies and enthalpies, where only changes could be measured, the Third Law enables us to measure absolute values of the entropy. The calculation simply consists of measuring the entropy changes in heating the compound from 0 K to whatever state it exists at 298 K.

Consider a gas at room temperature. In order to measure the *standard entropy*, S_{298}°, we need to determine the entropy changes in taking 1 mole of a compound through the following steps at 1 bar:

1. Heating from 0 K to T_m
$$\Delta S_1^\circ = \int_0^{T_m} \frac{\bar{c}_p(\text{solid})}{T} dT$$

2. Melting at T_m
$$\Delta S_2^\circ = \frac{\Delta H^{\text{fus}}}{T_m}$$

3. Heating from T_m to T_b
$$\Delta S_3^\circ = \int_{T_m}^{T_b} \frac{\bar{c}_p(\text{liquid})}{T} dT$$

4. Vaporization at T_b
$$\Delta S_4^\circ = \frac{\Delta H^{\text{vap}}}{T_b}$$

5. Heating from T_b to 298 K $\quad \Delta S_5^\circ = \int\limits_{T_b}^{298} \frac{\bar{c}_p(\text{vapour})}{T} \, dT$

$$S_{298}^\circ = \Delta S_1^\circ + \Delta S_2^\circ + \Delta S_3^\circ + \Delta S_4^\circ + \Delta S_5^\circ \tag{3.8}$$

For compounds that are liquid at 298 K and 1 bar, the final two steps are omitted and T_b is replaced by 298 K in the integral for ΔS_3°. In this calculation it is assumed that the mean heat capacities give sufficient accuracy and also that there is only one solid phase. In compounds with more phases in the solid state, the additional entropy associated with the solid phase changes must also be taken into account.

In practice, the integral in ΔS_1° cannot be evaluated because of the zero term in the integration limits. Thus, this integral must, at least at very low temperatures, be evaluated graphically. A plot of c_p/T versus T is constructed and extrapolated to 0 K. The area under the curve is equivalent to the integral in eqn 3.8 and can then be determined as in Fig. 3.3.

S_{298}° values are sometimes called *absolute* or *Third Law* entropies. Considering the values in Table 3.1 it is clear that they are consistent with our definition of entropy. For example, the three-dimensional diamond lattice is more strongly bound and organized than the layer structure of graphite so that it has a lower value of S_{298}°. Solids with more complex structures, particularly those with large numbers of atoms in the molecules, such as the sugars sucrose and glucose, have higher S_{298}° since there are a large number of ways of arranging the atoms in the molecule, and the molecules in the solid structures. S_{298}° can be thought of as increasing with the molecular complexity of the compound. Also as expected, in general, gases have higher S_{298}° than liquids which are in turn higher than solids (for molecules containing about the same number of atoms). This is best seen by considering the values for ice, water, and water vapour in Table 3.1.

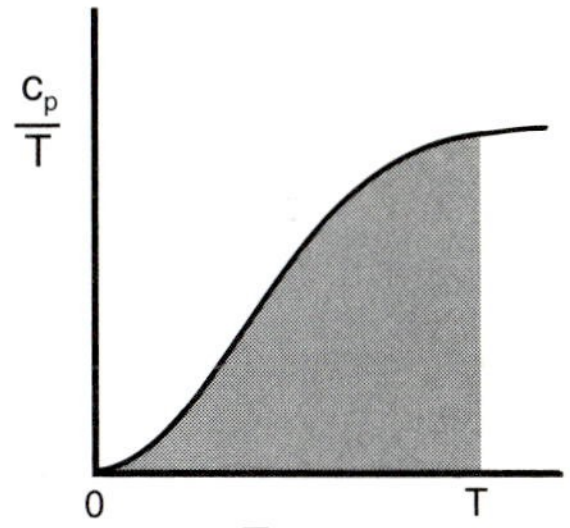

Fig. 3.4 Graphical determination of absolute entropies. The shaded area is equivalent to the standard entropy at temperature *T*.

3.4 Entropy changes in chemical reactions

In the same way as with enthalpy, we define the standard entropy of reaction as the difference in standard entropy of the products from that of the reactants, i.e.

$$\Delta S_{298}^\circ (\text{reaction}) = \sum v_i \, S_{298}^\circ(\text{products}) - \sum v_i \, S_{298}^\circ(\text{reactants}) \tag{3.9}$$

with v_i again signifying the stoichiometry. So, applying this to our general reaction,

$$\alpha \, A + \beta \, B \rightarrow \gamma \, C + \delta \, D$$

$$\Delta S_{298}^\circ = [\gamma \, S_{298}^\circ(C) + \delta \, S_{298}^\circ(D)] - [\alpha \, S_{298}^\circ(A) + \beta \, S_{298}^\circ(B)] \tag{3.10}$$

This gives the entropy change for converting 1 mole of reactants into 1 mole of products (or whatever the stoichiometry is) under standard pressure at 298 K. The entropy change at other temperatures can be calculated by applying eqn 3.6 in a manner analogous to the method used for enthalpy changes

$$\Delta S_{T_2} = \Delta S_{T_1} + \int\limits_{T_1}^{T_2} \frac{\Delta c_p}{T} \, dT \tag{3.11}$$

Exercise: Derive the equivalent equation for eqn 3.11 where c_p for each component is known as a power series as in eqn 2.11

with Δc_p again defined by eqn 2.10. Using mean heat capacities, this leads to

$$\Delta S_{T_2} = \Delta S_{T_1} + \Delta \bar{c}_p \ln \left(\frac{T_2}{T_1} \right) \tag{3.12}$$

Equation 3.12 allows calculation of entropy changes during reactions at any temperature, as demonstrated in Example 3.2.

Note that S_{298}°, or indeed the entropy of a compound under any other conditions, S_T, *must* be positive since it must have greater disorder than a perfect crystal at 0 K. However, ΔS_{298}° or ΔS_T can be positive or negative since they refer to *differences* in entropy between sets of conditions. It is important to keep this distinction in mind. Further examples of the calculation of entropy changes for reactions under various conditions will be presented in the following chapter.

Example 3.2   Using data in Appendix 1, calculate the entropy changes at 25 °C and at 750 °C for the following reactions:

   (a) $C_{(graph)} + O_{2(g)} \rightarrow CO_{2(g)}$
   (b) $N_{2(g)} + 3H_{2(g)} \rightarrow 2NH_{3(g)}$
   (c) $C_2H_5OH_{(l)} + CH_3COOH_{(l)} \rightarrow CH_3COOC_2H_{5(l)} + H_2O_{(l)}$

ΔS_{298}° can be calculated using eqn 3.9 and the variation at higher temperature using eqn 3.12. Appropriate data are taken from Appendix 1.

(a) $\Delta S_{298}^{\circ} = [S_{298}^{\circ}(CO_2)] - [S_{298}^{\circ}(C) + S_{298}^{\circ}(O_2)]$

$\quad = [213.74] - [5.74 + 205.14] \text{ J K}^{-1} \text{ mol}^{-1}$

$\quad = 2.86 \text{ J K}^{-1} \text{ mol}^{-1}$

$\Delta c_p = [\bar{c}_p(CO_2)] - [\bar{c}_p(C) + \bar{c}_p(O_2)] = [37.1] - [8.5 + 29.4]$

$\quad = -0.8 \text{ J K}^{-1} \text{ mol}^{-1}$

$\Delta S_{1023}^{\circ} = 2.86 + (-0.8) \ln (1023/298) = -1.87 \text{ J K}^{-1} \text{ mol}^{-1}$

(b) $\Delta S_{298}^{\circ} = [2 \times S_{298}^{\circ}(NH_3)] - [S_{298}^{\circ}(N_2) + 3S_{298}^{\circ}(H_2)]$

$\quad = (2 \times 192.45) - [191.61 + (3 \times 130.68)] \text{ J K}^{-1} \text{ mol}^{-1}$

$\quad = -198.75 \text{ J K}^{-1} \text{ mol}^{-1}$

$\Delta c_p = [2 \times \bar{c}_p(NH_3)] - [\bar{c}_p(N_2) + 3 \times \bar{c}_p(H_2)]$

$\quad = -38.2 \text{ J K}^{-1} \text{ mol}^{-1}$

$\Delta S_{1023}^{\circ} = -198.75 + (-38.2) \ln (1023/298)$

$\quad = -245.87 \text{ J K}^{-1} \text{ mol}^{-1}$

(c) $\Delta S_{298}^{\circ} = [S_{298}^{\circ}(H_2O) + S_{298}^{\circ}(CH_3COOC_2H_5]$

$\quad\quad\quad\quad - [S_{298}^{\circ}(CH_3COOC_2H_5N_2) + S_{298}^{\circ}(C_2H_5OH)]$

$\quad = [69.9 \times 259.4] - [159.8 + 160.7] \text{ J K}^{-1} \text{ mol}^{-1}$

$\quad = 8.8 \text{ J K}^{-1} \text{ mol}^{-1}$

$$\Delta c_p = (75.3 + 170.1) - (124.3 + 111.5) = 9.6 \text{ J K}^{-1} \text{ mol}^{-1}$$
$$\Delta S_{1023}^\circ = 8.8 + (9.6) \ln (1023/298) = 20.6 \text{ J K}^{-1} \text{ mol}^{-1}$$

The entropy change in (a) is relatively small since there are the same number of moles of gas in the reactants and the products. In (b), the number of moles of gas is halved, accounting for the large negative value of ΔS. In reaction (c), all the components are in the liquid phase so that again the change in entropy is relatively small.

Entropy as a predictor of chemical reactivity

Let us consider the entropy change for a familiar chemical reaction.

$$2H_{2(g)} + O_{2(g)} \rightarrow 2H_2O_{(l)}$$

From the data in Appendix 1, ΔS° can be calculated at 298 K.

$$\begin{aligned}
\Delta S_{298}^\circ &= [2 \times S_{298}^\circ(H_2O)] - [2 \times S_{298}^\circ(H_2) + S_{298}^\circ(O_2)] \\
&= (2 \times 70) - (2 \times 131 + 205) \\
&= -327 \text{ J K}^{-1} \text{ mol}^{-1}
\end{aligned}$$

Hence, there is a large loss of entropy on forming water, as would be expected in a reaction involving three moles of disordered, high-entropy gas forming two moles of relatively ordered and hence lower entropy liquid.

However, this result seemingly presents us with a problem. We know that water is stable with respect to its elements at 298 K and that the above reaction occurs spontaneously—try holding a match to a hydrogen–oxygen mixture! But doesn't the negative entropy change contravene the Second Law of thermodynamics, which states that spontaneous processes *increase* entropy? Of course, it doesn't—there are no known exceptions to any of the laws of thermodynamics. The apparent problem arises from an imprecise application of the Second Law. We have only calculated the entropy change for the $H_2/O_2/H_2O$ *system*. The Second Law refers to the entropy of the universe increasing as a requirement for spontaneity.

$\Delta H_{f,298}^\circ$ for water is $-285.8 \text{ kJ mol}^{-1}$ so the enthalpy change for the above reaction as written (i.e. the system) is -571.6 kJ. Clearly,

$$\Delta H^\circ(\text{surroundings}) = -\Delta H^\circ(\text{system})$$

From eqn 3.2,

$$\begin{aligned}
\Delta S(\text{surroundings}) &= \Delta H^\circ(\text{surroundings})/T = -\Delta H^\circ(\text{system})/T \\
&= +571.6 \times 10^3/298 \text{ J mol}^{-1}/\text{K} \\
&= +1920 \text{ J K}^{-1} \text{ mol}^{-1}
\end{aligned}$$

Considering the total entropy change,

$$\begin{aligned}
\Delta S(\text{universe}) &= \Delta S(\text{system}) + \Delta S(\text{surroundings}) \\
&= -327 + 1920 \\
&= +1593 \text{ J K}^{-1} \text{ mol}^{-1}
\end{aligned}$$

Hence, there is an overall *increase* in the total entropy of the universe, in accord with what is expected from the Second Law.

3.5   Concluding remarks

The ideas introduced in this chapter are some of the most fundamental in physical chemistry. In the context of this book, the main use of entropy is in providing a framework for describing the degree of organization in a system.

It also gives us a method for predicting the course of chemical reactions (or other processes). However, it is not very convenient to have to consider the whole universe every time we want to study a particular reaction and it would be useful to focus attention only on the system that is of interest. The development of suitable methods for this will be the subject of the next chapter.

4 Free energy and equilibrium

The aim of this chapter (and in some ways the book as a whole) is to try to achieve an understanding of what factors govern chemical reactivity. We have already seen that there are essentially two driving forces which operate in chemical processes. First, we consider the *energy changes*—governed by the enthalpy or internal energy. Secondly, there is the tendency of systems to achieve greater disorder—correlated by the *change in entropy*. We have developed equations to calculate how each of these factors change during a reaction and are now in a position to consider the relation between them.

4.1 Free energy and reactivity

If we think about chemical reactions which we commonly encounter, most (but not all) reactions are exothermic with negative ΔH. It is also common experience that most reactions (but not all) lead to an increase in disorder in the system—positive ΔS as embodied in the Second Law. However, is there a way of combining these into a single function which will govern the viability of a process? Of course there is, and this is the *free energy*.

The idea of free energy follows directly from the Second Law of thermodynamics. We saw in the previous chapter that for a spontaneous process at constant temperature and pressure:

$$\Delta S(\text{universe}) = \Delta S(\text{system}) + \Delta S(\text{surroundings}) > 0$$

Again drawing on our discussion in the previous chapter, this can be rewritten as

$$\Delta S(\text{universe}) = \Delta S(\text{system}) + [\Delta H(\text{surroundings})/T] > 0$$

However, we want to focus on the system without the encumbrance of considering the rest of the universe every time we study a reaction. As shown in Fig. 4.1, the enthalpy change for the surroundings must be the reverse of that for the system so that $\Delta H(\text{surroundings}) = -\Delta H(\text{system})$. Hence,

$$\Delta S(\text{system}) + [-\Delta H(\text{system})/T] > 0$$

We now have terms referring only to the system so we can drop the labels and rearrange the equation:

$$\Delta H - T\,\Delta S < 0 \tag{4.1}$$

where everything now refers to a property of the system.

The value of $H - TS$ has such a special significance that it is given its own name, the *Gibbs function* or *Gibbs free energy*, G. An exact definition of G is therefore given by:

$$G = H - TS \tag{4.2}$$

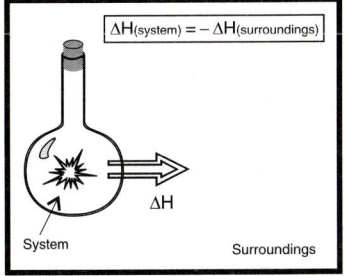

Fig. 4.1 Enthalpy changes for the system and the surroundings.

Since it is a combination of two state functions, G must also be a state function. Again we therefore use the convention that ΔG is the difference in Gibbs free energy between the final and initial states. Hence,

$$\Delta G = G_{(\text{final})} - G_{(\text{initial})} = \Delta H - T\Delta S \tag{4.3}$$

It is clear from eqn 4.1 that, for a spontaneous process at constant temperature and pressure, there must be a lowering of the Gibbs free energy or

$$(\Delta G)_{p,T} < 0 \tag{4.4}$$

Equation 4.4 is perhaps the most important result of any that will be presented in this book. It is the fundamental basis for the explanation of chemical reactivity, chemical equilibrium, and the phase behaviour of compounds.

Note: The proof that ΔG represents the maximum non-pV work which can be done by a system can be found in the books listed in the Glossary at the end of this Primer.

Although we will not prove it here, it can be shown that ΔG also represents the maximum amount of work that a system can do at constant temperature and pressure after any p–V work has been accounted for. (This will be particularly important in, for example, electrochemical cells and biochemical systems where expansion work does not feature.) It is also the origin of the name 'free' energy—it represents the energy of a system which is 'free' to drive a chemical process. The more modern name for G is the Gibbs function but the older terminology is still widely used and will be retained here.

The above arguments have been developed in terms of enthalpy changes at constant pressure, but an analogous argument can be made in terms of internal energy changes occurring at constant volume. In this case, we define the *Helmholtz free energy*, A, as $A = U - TS$. Thus, spontaneous changes at constant volume have $\Delta A < 0$. The remainder of this book will concentrate on applications of the Gibbs free energy since most chemical reactions in the laboratory are conducted at constant pressure. However, ΔA can be useful in a number of situations. It is commonly used in discussions of solid and condensed phase processes where the changes in volume can be negligibly small.

Note: The Helmholtz free energy is commonly used in consideration of geochemical problems where huge pressure changes can be involved but the volume of a system does not change markedly.

Looking at the definition of the Gibbs free energy change embodied by eqn 4.3, we can explain the general observations on reactions which were stated at the beginning of this chapter. At low temperatures, the $T\Delta S$ term will be relatively small compared with ΔH so that ΔG is insensitive to the sign of the entropy change. (Don't forget, we work in kelvin so T can never be negative.) The sign of ΔG is therefore determined by the sign of ΔH, as shown schematically in Fig. 4.2. At low temperatures, the entropic contribution can be negative or positive but the Gibbs free energy change will be negative. Hence, for a spontaneous reaction, we need to have negative ΔH or an exothermic reaction. Conversely, at high temperatures the $(-T\Delta S)$ term will be dominant so that a positive ΔS, hence a disordering process, is needed to give a negative ΔG.

In general, we can conclude that *if a process or reaction results in a lowering of the free energy, it will be spontaneous*, i.e. that it will be favourable and CAN happen. It does not mean that it WILL happen. For example, we will see shortly that the reaction of hydrogen and oxygen gases to form water at 1 bar and 25 °C has a highly negative change in free energy.

$$2H_{2(g)} + O_{2(g)} \rightarrow 2H_2O_{(l)} \quad \Delta G^\circ = -273.2 \text{ kJ mol}^{-1}$$

However, a mixture of hydrogen and oxygen gases in a container will exist with no reaction for a long time until started by a catalyst or source of ignition,

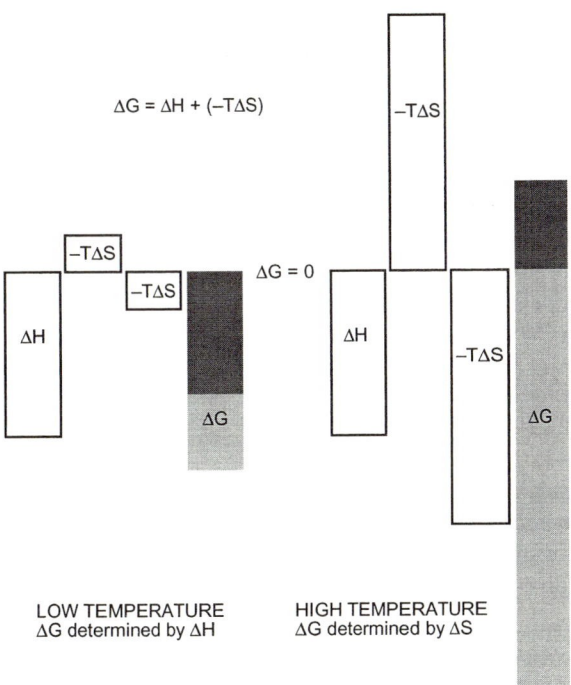

Fig. 4.2 Contributions to the change in free energy. The shaded areas show the resultant free energy changes. At the extremes of temperature, the dominant factor is sufficient to overcome the other, whatever the sign of the latter.

whereupon it proceeds very well. The explanation for this is that the reaction is *thermodynamically* very favourable but occurs infinitely slowly. It is *kinetically* unfavourable as it has a high activation energy.

If we can calculate the value of ΔG for any reaction or process in which we are interested, we can make some predictions on its feasibility. It is this that will concern us for the remainder of this chapter.

4.2 Gibbs free energies of formation

In the same way as we defined standard enthalpies of formation in Chapter 2, we can also define standard Gibbs free energies of formation, $\Delta G_{f,298}^{\circ}$

> $\Delta G_{f,298}^{\circ}$ *is the change in Gibbs free energy when 1 mole of a compound is formed at 1 bar and 298 K from its elements in their standard states.*

Tables of $\Delta G_{f,298}^{\circ}$ have been drawn up and some data are given in Appendix 1 and Table 4.1. These values can be used in a manner analogous to $\Delta H_{f,298}^{\circ}$ to calculate the change in free energy during a reaction (see Example 4.1). Since G is made up from two state functions, S and H, it must itself be a state function. Thus we can use the (hopefully familiar by now) form

$$\Delta G_{298}^{\circ}(\text{reaction}) = \sum v_i \Delta G_{f,298}^{\circ}(\text{products}) - \sum v_i \Delta G_{f,298}^{\circ}(\text{reactants}) \quad (4.5)$$

It is apparent that, as in the case of ΔH_f°, the value of ΔG_f° must be zero for a pure element in its standard state. We should be careful about drawing too

Table 4.1 Standard Gibbs energy of formation, $\Delta G_{f,298}^{\circ}$, for selected compounds, in kJ mol^{-1}

$C_{(graphite)}$	0
$C_{(diamond)}$	+2.9
$CO_{(g)}$	−137.2
$CO_{2(g)}$	−394.4
$CH_{4(g)}$	−50.7
$C_2H_{6(g)}$	−32.8
$F_{2(g)}$	0
$HF_{(g)}$	−271.2
$H_{2(g)}$	0
$H_2O_{(l)}$	−237.1
$H_2O_{(g)}$	−228.6
$N_{2(g)}$	0
$NH_{3(g)}$	−16.5
$NO_{2(g)}$	+51.3
$N_2O_{4(g)}$	+97.9
$O_{2(g)}$	0
$O_{3(g)}$	+163.0

Using eqns 2.3, 3.9 and 4.3.

$\Delta H^{\circ}_{298} = [0 + (-393.5)] - (-601.7) + (-110.5)] = +318.7 \text{ kJ mol}^{-1}.$

$\Delta S^{\circ}_{298} = [32.7 + 213.7] - [26.9 + 197.7] \qquad = 21.8 \text{ J mol}^{-1} \text{ K}^{-1}.$

$\Delta G^{\circ}_{298} = \Delta H^{\circ}_{298} - T\Delta S^{\circ}_{298} = 318.7 - (298.15)(21.8 \times 10^{-3}) \text{ kJ mol}^{-1}.$

$\qquad = +312.2 \text{ kJ mol}^{-1}$

To convert ΔH°_{298} and ΔS°_{298} to the higher temperature, we need Δc_p.

$\Delta c_p = (24.9 + 37.1) - (37.2 + 29.1) \qquad = -4.3 \text{ J mol}^{-1} \text{ K}^{-1}.$

$\Delta H^{\circ}_{1298} = \Delta H^{\circ}_{298} + \Delta c_p(1298 - 298) = 318.7 + (-4.3 \times 10^{-3})(1000)$

$\qquad = 314.4 \text{ kJ mol}^{-1}.$

$\Delta S^{\circ}_{1298} = \Delta S^{\circ}_{298} + \Delta c_p \ln (1298/298) = 21.8 + (-4.3)(1.471)$

$\qquad = 15.47 \text{ J mol}^{-1} \text{ K}^{-1}.$

$\Delta G^{\circ}_{1298} = \Delta H^{\circ}_{1298} - T\Delta S^{\circ}_{1298} = 314.4 - (1298.15)(15.47 \times 10^{-3}) \text{ kJ mol}^{-1}.$

$\qquad = +294.3 \text{ kJ mol}^{-1}.$

Reduction of a metal oxide with CO is sometimes used to recover the metal from an ore. In this case, with such a positive change of Gibbs' free energy, this is clearly not feasible

Variation of *G* with pressure

This can be obtained by an alternative approach to eqn 4.7, by varying the pressure at fixed temperature so that $dT = 0$.

$$\left(\frac{dG}{dG}\right)_T = V \qquad (4.10)$$

As noted previously (Section 2.4), the molar volumes of solids and liquids, and particularly their changes during reactions, are relatively small and so there is usually little noticeable change of G or ΔG with pressure. A more useful example is given by considering an ideal gas. If G is here defined as the molar Gibbs free energy, i.e. the Gibbs free energy per mole of pure component,

$$\left(\frac{dG}{dp}\right)_T = V = \frac{nRT}{p}$$

Hence, if the free energy changes from G_1 to G_2 when the pressure changes from p_1 to p_2,

$$\int_{G_1}^{G_2} dG = \int_{p_1}^{p_2} \frac{nRT}{p} dp = nRT \int_{p_1}^{p_2} \frac{1}{p} dp \qquad (4.11)$$

which leads to $\qquad G_2 - G_1 = nRT \ln \left(\frac{p_2}{p_1}\right)$

A particularly important form of this equation is to recall that G° is the free energy at the standard pressure of $p^{\circ} = 1$ bar. Hence,

$$G = G^{\circ} + nRT \ln (p/p^{\circ}) = G^{\circ} + nRT \ln (p/1 \text{ bar}) \qquad (4.12)$$

Note: Although we will always assume that gases behave ideally in this primer, it is important to realize that real gases will deviate from eqn 4.12. To account for this, we define a *fugacity* or 'effective pressure', f, where $f = \phi(p)$. ϕ is a factor which describes how the behaviour deviates from the ideal model where $\phi = 1$. The ideal model is an adequate description of gas behaviour under most conditions but ϕ accounts for the differences due to the intermolecular interactions at high pressures.

Equation 4.12 is often written as '$G = G^\circ + nRT \ln (p)$'. This is correct as long as the pressure is measured in units of bar, but it should always be remembered that the quantity in brackets must be a dimensionless quantity and is thus the ratio of the pressure to a standard value.

The thermodynamic activity

In order to achieve a consistent representation of the concentration dependence of free energy, we define a new parameter, the *activity* or 'effective concentration' of the component. Thus we can use the general equation

$$G = G^\circ + nRT \ln a \qquad (4.13)$$

with the activity, a, and standard states appropriately defined.

$$\text{Ideal gases}: \quad a = p/p^\circ \quad p^\circ = 1 \text{ bar} \qquad (4.14)$$

Since the effect of pressure on solids and liquids is negligible and, by definition $G = G^\circ$ for a pure liquid or solid, then it follows that the activity of either of these states must be unity. Thus,

$$a_{(\text{pure solid})} = a_{(\text{pure liquid})} = 1 \qquad (4.15)$$

We will see other applications of the thermodynamic activity when we discuss phase equilibrium in Chapter 5.

Variation of *G* with concentration

Of course, not all chemical systems exist in the gas phase so that we need to have a measure of how G varies in condensed phase systems. Although eqn 4.11 was derived from consideration of the pressure of an ideal gas, it expresses the dependence on concentration (i.e. number of moles per unit volume) since $p = RT (n/V)$. It also predicts a straightforward dependence of G on composition.

For convenience, we can adopt an expression of the same form for reactions in solution by considering the *concentration* of the component in which we are interested. Here, we begin by using *molar* concentrations so that we adopt as our standard state a concentration of 1 mol dm^{-3}. The molar concentration of a compound i is often represented as $[i]$ so we can write

$$G_i = G_i^\circ + nRT \ln ([i]/1 \text{ mol dm}^{-3}) \qquad (4.16)$$
$$= G_i^\circ + nRT \ln [i] \qquad (4.17)$$

Note: The use of molar concentrations can be regarded as a sort of 'ideal' approximation which is valid at low concentrations. The way that this is overcome can be seen in the more advanced texts listed in the Further Reading.

Given the equations and ideas developed so far, it is now possible to take some data (readily available in tables) and calculate some properties of real reactions. However, first we need to investigate further the influence that the Gibbs free energy change has on the extent of chemical reactions.

4.4 Free energy and equilibrium

So far, we have used ΔG simply to predict the spontaneity or otherwise of the reaction. ΔG is in fact much more useful than this. The free energy will change throughout a reaction depending on the proportions of reactants and products (we will see how to calculate this shortly) but will reach a minimum value at some composition (which may effectively be 100% reactant—no reaction—or 100% product—complete reaction—or any value in between). At

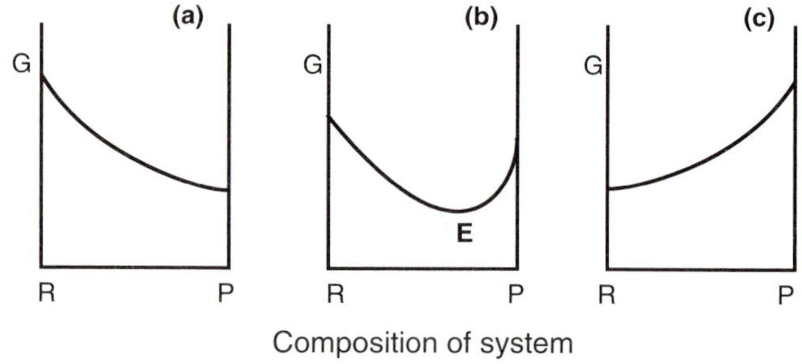

Fig. 4.3   Schematic representation of Gibbs free energy changes during reactions.
R = reactants; P = products. (a) Reaction goes to completion; (b) reaction comes to equilibrium
at composition, E; (c) reaction does not go at all.

this minimum, the reaction will have no tendency to go anywhere and so will
stop. It has reached equilibrium. This can be illustrated by the plots shown in
Fig. 4.3.

For the system in (a), the Gibbs free energy of the products, P, is clearly
lower than those of the reactants, R. ΔG is therefore negative and the reaction
proceeds in a forward direction. The converse is true for the system shown in
(c). For the system shown in (b), ΔG would be negative if the system started at
either 100% P or 100% R so that the reaction would proceed until point E. If
the composition were changed from this point, the result would be an increase
in the free energy. Thus, the reaction cannot change from this composition—*it
has come to equilibrium*. This schematic approach allows us to define one of
the fundamental statements of chemistry

*A system will come to equilibrium when it has reached its minimum free
energy.*

To make further progress, we need to have a quantitative measure of the
composition of a reaction mixture. A number of approaches are available but
one of the most straightforward makes use of the familiar *law of mass action*
to define the reaction quotient, Q.

For our generalized reaction, $\alpha\,A + \beta\,B \rightarrow \gamma\,C + \delta\,D$, the reaction
quotient, Q, is defined by

$$Q = \frac{(a_C)^\gamma (a_D)^\delta}{(a_A)^\alpha (a_B)^\beta} \qquad (4.18)$$

Equation 4.18 is completely general and to relate the activity to measurable
parameters we make use of the definitions in the previous section.

When the system reaches equilibrium,

$$Q_{\text{eqm}} = \frac{(a_C)^\gamma_{\text{eqm}} (a_D)^\delta_{\text{eqm}}}{(a_A)^\alpha_{\text{eqm}} (a_B)^\beta_{\text{eqm}}} = K_{\text{eqm}} \qquad (4.19)$$

where the subscript 'eqm' indicates that the activities are measured after the
reaction has come to equilibrium. K_{eqm} is therefore the *equilibrium constant*
for the reaction.

Different types of equilibrium constant can be defined depending on the
reactions involved. If K_{eqm} is calculated from gas pressures, the equilibrium

Mathematical note: An alternative way of
writing eqn 4.19 is

$$Q = \frac{\Pi(a_{\text{products}})^\nu}{\Pi(a_{\text{reactants}})^\nu}$$

where ν is the stoichiometric coefficient.
The symbol Π may not be familiar. It is
shorthand for 'the product of' in the same
way as $\sum$ represents 'the sum of'. Thus

$$\Pi_1^5(a_i) = a_1 \times a_2 \times a_3 \times a_4 \times a_5$$

constant is K_p; if calculated from solution concentrations then it is K_c. This is best illustrated by taking examples.

$$N_{2(g)} + 3H_{2(g)} \leftrightarrow 2NH_{3(g)} \quad K_p = \frac{(p_{NH_3})_{eqm}^2}{(p_{H_2})_{eqm}^3 (p_{N_2})_{eqm}}$$

$$CaO_{(s)} + CO_{2(g)} \leftrightarrow CaCO_{3(s)}$$

$$K_p = \frac{(a_{CaCO_3})}{(p_{CO_2})_{eqm}(a_{CaO})} = \frac{1}{(p_{CO_2})_{eqm}(1)} = \frac{1}{(p_{CO_2})_{eqm}}$$

$$HCOOH_{(aq)} + C_2H_5OH_{(aq)} \leftrightarrow HCOOC_2H_{5(aq)} \leftrightarrow H_2O_{(l)}$$

$$K_c = \frac{[HCOOC_2H_5]_{eqm}[H_2O]_{eqm}}{[HCOOH]_{eqm}[C_2H_5OH]_{eqm}}$$

$$CH_3COOH_{(aq)} \leftrightarrow H^+_{(aq)} + CH_3COO^-_{(aq)} \quad K_c = \frac{[CH_3COO^-]_{(eqm)}[H^+]_{(eqm)}}{[CH_3COOH]_{(eqm)}}$$

In these examples, p indicates the partial pressure of a gas measured in bar, (aq) indicates the component is in aqueous solution, (eqm) signifies that the reaction has reached equilibrium, and [ ] indicates the solution concentration in mol dm^{-3}.

The definition of Q and K_{eqm} allow us to predict the direction of a chemical reaction. Drawing together the material from the last two sections, we are now in a position to say that, if we know ΔG and/or Q for a reaction or process, then

Note: Even though we have written the pressures or concentrations in the expressions for the equilibrium constants, it should be remembered that they are in fact ratios, e.g. $p/1$ bar, $c/$mol dm^{-3}, etc. The consequence of this is that the equilibrium constant is dimensionless and has no units.

If $\Delta G < 0$ or $Q < K_{eqm}$ the forward reaction proceeds
If $\Delta G > 0$ or $Q > K_{eqm}$ the reverse reaction proceeds
If $\Delta G = 0$ or $Q = K_{eqm}$ the system is at equilibrium

The next step in the story is to define the quantitative relationship between the free energy change for a reaction and its equilibrium position.

4.5 Relation between free energy and the equilibrium constant

Consider again our general reaction. For convenience here, we will ignore the stoichiometry and assume that one molecule of each species is involved in the reaction:

$$A + B \rightarrow C + D$$
$$\Delta G(\text{reaction}) = G(\text{products}) - G(\text{reactants}) = (G_C + G_D) - (G_A + G_B)$$

We know from eqn 4.13 that $G_A = G_A^\circ + RT \ln a_A$, and similarly for the other components of the reaction. Even though this was derived for a single gas or a single component in solution, let us begin by assuming that it can also be applied when the component appears in a mixture.

$$\Delta G(\text{reaction}) = [(G_C^\circ + RT \ln a_C) + (G_D^\circ + RT \ln a_D)] \tag{4.20}$$
$$- [(G_A^\circ + RT \ln a_A) + (G_B^\circ + RT \ln a_B)]$$
$$= [G_C^\circ + G_D^\circ) - (G_A^\circ + G_B^\circ)] + RT[(\ln a_C + \ln a_D) - (\ln a_A + \ln a_B)]$$
or $$\Delta G(\text{reaction}) = \Delta G^\circ(\text{reaction}) + RT \ln Q \tag{4.21}$$

From above, when the system reaches equilibrium, the free energy will be at a minimum so that $\Delta G(\text{reaction}) = 0$ and $Q = K_{\text{eqm}}$. Hence,

$$0 = \Delta G^\circ(\text{reaction}) + RT \ln K_{\text{eqm}}$$

$$K_{\text{eqm}} = \exp\left(\frac{-\Delta G^\circ}{RT}\right) \tag{4.22}$$

Exercise: Take the general reaction described by eqn 2.4 and prove that when the stoichiometry is taken into account, eqns 4.21 and 4.22 are valid.

This is another very important equation in describing the physical chemistry of reactions. The derivation is specific for the reaction listed above. However, the same method applied to *any* reaction with *any* stoichiometry leads to the same result. It is worthwhile stressing one feature of eqn 4.21. At equilibrium, ΔG for the reaction under the equilibrium conditions will be zero; ΔG°, which is defined under standard conditions, will not, in general, equal zero.

Gibbs free energy and reactivity

In Section 4.1, we used the sign of the free energy change to determine whether a reaction was spontaneous. The equations outlined in this section allow us to embark on a rather more detailed discussion of the role of free energy.

Around room temperature—say 298 K—we know that if ΔG° for a reaction is negative, $K_{\text{eqm}} > 1$ so that the forward reaction will take place and the products will be present in larger concentration than the reactants. Can we quantify this?

In our ideal world, let us define a complete reaction as one which gives a yield of 99.99% products. Most synthetic chemists would hold this as an unattainable goal but in thermodynamics we can set our standards somewhat higher!

$$K_{\text{eqm}} = \frac{\text{Amount of products}}{\text{Amount of reactants}} = \frac{99.99}{0.01} = 9999 \approx 10^4$$

For $K_{\text{eqm}} = 10^4$ at 298 K, ΔG° must be -22.8 kJ mol^{-1} so any reaction with a more negative value will go to completion. The converse argument shows that any reaction with ΔG° more positive than $+22.8$ kJ mol^{-1} will not occur to any noticeable extent. Between these 'limits', a variety of behaviour results. This can be expressed graphically as in Fig. 4.4.

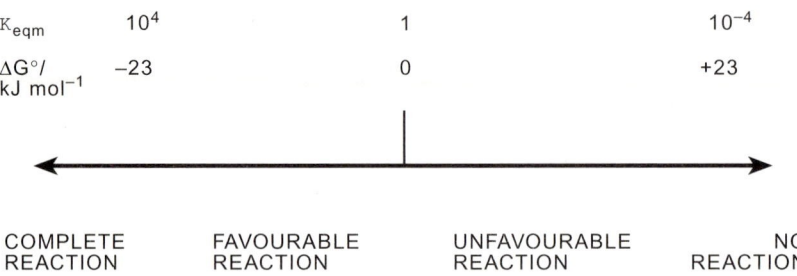

Fig. 4.4   Schematic representation of the relation between Gibbs free energy and reactivity.

Exercise: Calculate the value of ΔG° required for a complete reaction at 1000 K.

It should be noted that the 'limits' in Fig. 4.4 are temperature dependent because of the $-T\Delta S^\circ$ contribution to the Gibbs free energy.

Pause for breath—a recap!

Having reached this important point, let us recap what we have achieved. With the methods we have developed so far, we can take data from the literature in terms of standard entropies, $\Delta H^{\circ}_{f,298}$ and heat capacities for the components involved in a reaction, calculate ΔH° and ΔS° at any temperature, calculate ΔG° and hence the equilibrium constant. This allows us to predict the course of the reaction, and the position of equilibrium clearly gives us the maximum possible yield of products. It is in performing these relatively straightforward calculations that the immense power of thermodynamics lies.

Example 4.3 Silver carbonate decomposes on heating. Calculate the equilibrium constant at 110 °C for the reaction.

$$Ag_2CO_{3(s)} \rightarrow Ag_2O_{(s)} + CO_{2(g)}$$

	$Ag_2CO_{3(s)}$	$Ag_2O_{(s)}$	$CO_{2(g)}$
$\Delta H^{\circ}/\text{kJ mol}^{-1}$:	−501.4	−29.07	−393.5
$\Delta S^{\circ}/\text{J K}^{-1}\text{mol}^{-1}$:	167.3	121.7	213.7
$c_p/\text{J K}^{-1}\text{ mol}^{-1}$:	109.6	68.6	37.1

Proceeding along similar lines to Example 4.3,

$\Delta H^{\circ}_{298} = [(-29.07) + (-393.5)] - [-501.4)] = +79.03 \text{ kJ mol}^{-1}$.

$\Delta S^{\circ}_{298} = (121.7 + 213.7) - (167.3) \qquad = 168.0 \text{ J mol}^{-1} \text{ K}^{-1}$.

$\Delta c_p = (68.6 + 37.1) - (109.6) \qquad = -3.9 \text{ J mol}^{-1} \text{ K}^{-1}$.

$\Delta H^{\circ}_{383} = 79.03 + (-3.9 \times 10^{-3})(383.15 - 298.15) = 78.70 \text{ kJ mol}^{-1}$

$\Delta S^{\circ}_{383} = 168.0 + (-3.9) \ln (383.15/298.15) = 167.0 \text{ J mol}^{-1} \text{ K}^{-1}$.

$\Delta G^{\circ}_{383} = 78.7 - (383.15)(167.0 \times 10^{-3}) = 14.71 \text{ kJ mol}^{-1}$.

We now use eqn 4.23 to find the equilibrium constant.

$$K_p = \exp\left(\frac{-\Delta G^{\circ}}{RT}\right) = \exp\left(\frac{-14.71 \times 10^3}{(8.314)(383.15)}\right) = 9.87 \times 10^{-3}$$

To illustrate how this result can be used, say a sample of $Ag_2CO_{3(s)}$ is to be dried in an oven. The reaction quotient, Q, for this reaction is simply $p(CO_2)$. If the conditions in the oven are such that the partial pressure of CO_2 is $> 9.9 \times 10^{-3}$ (or ~1% by volume) then the silver carbonate will not decompose.

4.6 Chemical potential

Before proceeding, we should discuss the approximation made in deriving eqn 4.24, which is that eqn 4.13 can be applied when a component is present in a mixture, i.e.

$$G_A = G^{\circ} + RT \ln a_A$$

This is equivalent to saying that the free energy of the component is independent of all the other components of the mixture. This is reasonable for ideal gases but not for other systems.

Let us think about 1 mole of water. Is it reasonable to assume that the free energy change will be the same if it is added to a large amount of water, or

Note: In fact, this somewhat more complex treatment ends up with exactly the same result in terms of the free energy and equilibrium constant as that in Section 4.4 and therefore could be omitted. However, it is more logically correct and worthwhile working through!

Mathematical note: The function given by eqn 4.25 is an example of a partial derivative. It occurs in a system of several variables when all others are kept constant and the differential with respect to the one of interest obtained.

benzene, or ethanol, or sulfuric acid? Clearly not. What is of interest to us here is how the free energy changes when a component is added to the system with all other conditions and amounts of other components kept equal. Mathematically, this is expressed by the *partial differential* of G with respect to that component.

$$\bar{G}_A = \left(\frac{\partial G}{\partial n_A}\right)_{P,T,n_B} \tag{4.23}$$

$\bar{G}_A$ is the *partial molar free energy* of component A and represents the free energy change when 1 mole of A is added to a system at constant conditions. To avoid this lengthy name (and to stress its importance which will become apparent shortly) it is called the *chemical potential* and given the symbol μ. For a pure substance, μ_A corresponds to the molar Gibbs free energy.

If we have a large amount of a substance, B, and we add an amount dn_A moles of a second substance, A, then the free energy change at constant temperature and pressure, dG, is

$$dG = \mu_A \, dn_A$$

For a multicomponent system, this can be generalized to

$$(dG)_{p,T} = \sum_i \mu_i \, dn_i \tag{4.24}$$

The fundamental equation of thermodynamics introduced above as eqn 4.7 can therefore be more completely written to account for changes in all parameters as:

$$dG = V \, dP - S \, dT + \sum_i \mu_i \, dn_i \tag{4.25}$$

To illustrate the importance of chemical potential, we start with eqn 4.12, $G = G° + nRT \ln (p/p°)$. If we consider solely component A in the mixture, then

$$\left(\frac{\partial G}{\partial n_A}\right) = \left(\frac{\partial G°}{\partial n_A}\right) + RT \ln\left(\frac{p_A}{p_A°}\right)$$

or

$$\mu_A = \mu_A° + RT \ln (p_A/p_A°) \tag{4.26}$$

where $\mu_A°$ is the standard chemical potential at the standard partial pressure of $p_A° = 1$ bar. This is identical with the free energy of 1 mole of pure A.

Now let us return to a model reaction A $\rightarrow$ B. If the reaction proceeds so that dn moles of A react forming dn moles of B, then the change in free energy, dG, is given by

$$dG = (-dn \, \mu_A) + dn \, \mu_B \tag{4.27}$$

$$= -dn[\mu_A° + RT \ln (p_A/1 \text{ bar})] + dn[\mu_B° + RT \ln (p_B/1 \text{ bar})]$$

$$dG/dn = [(\mu_B° - \mu_A°) + RT \ln (p_B/p_A)] = \Delta G° + RT \ln (p_B/p_A)$$

Exercise: Prove the above for the more complex model reaction which we have used in the previous sections, eqn 2.4.

At equilibrium, $dG/dn = 0$ and $(p_B/p_A) = K_p$. Hence,

$$0 = \Delta G° + RT \ln (p_B/p_A) = \Delta G° + RT \, K_p$$

or

$$\Delta G° = -RT \, K_p$$

which is the equivalent to eqn 4.22.

Although this treatment leads to eqn 4.22, which is important in itself, a further insight into the significance of chemical potentials can be gained by looking again at eqn 4.27. As we have seen before, $dG = 0$ when the reaction reaches equilibrium.

$$0 = [\mu_A(-dn)] + \mu_B dn \quad \text{or} \quad \mu_A dn = \mu_B dn \quad \mu_A = \mu_B$$

Without proving it for more complicated cases, this can be stated as an example of *a general requirement for chemical equilibrium.*

> *A reaction comes to equilibrium at constant temperature and pressure under conditions where the chemical potential of the reactants equals that of the products.*

You may be wondering 'why the name—chemical potential?'. In fact, by analogy with other, perhaps more familiar situations, the reason for the name should become apparent. If a body is raised to a greater altitude, it has a greater 'driving force' to fall to earth—we say it has a higher *gravitational potential* and the body will spontaneously move from a position of high gravitational potential to one with a lower value. In an electric circuit, electrons move around the circuit under the influence of a voltage. A high voltage—or *electrical potential*—provides a greater driving force for the electrons to move from high potential to low. In a similar manner, the chemical potential can be thought of as the 'driving force' for a chemical process. As illustrated in Fig. 4.5, μ is equivalent to the slope of the Gibbs free energy *versus* composition curve. The steeper the curve, and hence higher μ, the greater the tendency for the system to move toward equilibrium where μ is zero.

Although this discussion has centred on chemical reactions, μ can be used in a variety of situations. We will meet it again in Chapter 5.

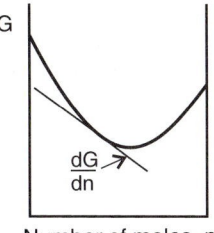

Fig. 4.5 Relation between chemical potential and free energy changes during a reaction.

4.7   Relating the equilibrium constant to experiment

To complete this section of 'reaction thermodynamics' we now simply (!) have to relate the equilibrium constant to the amounts of reactant and products in our reactions. Again, the easiest way to proceed is to take a particular reaction to act as an example.

Consider the gas phase reaction:

$$N_2O_{4(g)} \leftrightarrow 2\,NO_{2(g)}$$

If 1 mol of N_2O_4 is introduced into a container and allowed to reach equilibrium at a total gas pressure of P_{tot} bar, let us say that α moles of N_2O_4 will have reacted.

At equilibrium, there will be $(1 - \alpha)$ mol of N_2O_4 remaining and 2α mol of NO_2 will have been formed. The total number of moles of gas present is therefore $(1 - \alpha) + 2\alpha = (1 + \alpha)$ mol.

From the definition of the equilibrium constant,

Note: α is more completely defined as the fraction of moles of starting material that react and is often termed the degree of dissociation.

$$K_p = \frac{(p_{NO_2})^2}{(p_{N_2O_4})} = \frac{(x_{NO_2}P_{tot})^2}{(x_{N_2O_4}P_{tot})} = \frac{\left(\frac{2\alpha}{1+\alpha}P_{tot}\right)^2}{\left(\frac{1-\alpha}{1+\alpha}P_{tot}\right)}$$

Note: For gases acting ideally, the partial pressure, p, is given by Dalton's law:

p = mole fraction × total pressure

$= xP_{tot}$

The mole fraction, x, is given by:

$$x = \frac{\text{No. of moles of component}}{\text{Total no. of moles present}}$$

Note: You should select a number of examples from different texts and work through them to derive the relation between K_{eqm} and the extent of reaction. It is one of those areas where practice really does make perfect!

$$K_p = \frac{4\alpha^2 P_{tot}^2 (1 + \alpha)}{(1 - \alpha)(1 + \alpha)^2 P_{tot}} = \frac{4\alpha^2 P_{tot}}{(1 - \alpha)(1 + \alpha)}$$

or

$$K_p = \frac{4\alpha^2 P_{tot}}{(1 - \alpha^2)}$$

Hence, if K_p can be calculated from thermodynamic data, the value of α can be calculated at any pressure. From this, the maximum amount of product can be readily found.

The bad news here is that, although this method is extremely powerful, each different reaction stoichiometry leads to a different expression for K_p (or K_c as appropriate) and the resulting equations can be difficult to solve analytically. Although solution of the equations presents no challenge to numerical methods, particularly those available for computers, it is often useful to gain rapidly an approximation of the solution. One simple technique used in these cases is to assume that the degree of reaction is small so that $\alpha \ll 1$ and so $(1 - \alpha) \approx 1$. Thus for the reaction considered above,

$$K_p \approx 4\alpha^2 P_{tot}$$

and it is readily apparent that $\alpha \approx \sqrt{K_p/4\,P_{tot}}$.

Example 4.4   Formulate expressions in terms of the degree of dissociation for the equilibrium constants for the following reactions.

(a) $Ni(CO)_{4(g)} \leftrightarrow Ni_{(s)} + 4CO_{(g)}$
(b) $HgO_{(s)} \leftrightarrow Hg_{(l)} + O_{2(g)}$
(c) $CH_3COOH_{(aq)} \leftrightarrow CH_3COO^-_{(aq)} + H^+_{(aq)}$

(a) If we start with 1 mole $Ni(CO)_{4(g)}$ and α react, then at equilibrium there will be $1 - \alpha$ moles remaining, α moles of $Ni_{(s)}$ together with 4α moles of CO. When considering K_p, only gases contribute to the partial pressures. The total number of gas moles $= (1 - \alpha) + 4\alpha = (1 + 3\alpha)$

$$K_p = \frac{(1)(p_{CO})^4}{p_{Ni(CO)_4}} = \frac{(1)\left(\frac{4\alpha}{1+3\alpha}\right)^4 P_{tot}^4}{\left(\frac{1-\alpha}{1+3\alpha}\right)P_{tot}} = \frac{(4\alpha)^4 (1 + 3\alpha)P_{tot}^3}{(1 - \alpha)(1 + 3\alpha)^4} = \frac{256\,\alpha^4 P_{tot}^3}{(1 - \alpha)(1 + 3\alpha)^3}$$

(b) The only gas involved is O_2. Thus, since the activity of solids is 1 and the only gas present is oxygen,

$$K_p = \frac{(a_{Hg})(p_{O_2})}{(a_{HgO})} = \frac{(1)(p_{O_2})}{(1)} = P_{tot}$$

(c) If the initial concentration is c mol dm^{-3} and α is the fraction that reacts, at equilibrium there will be (αc) mol dm^{-3} of H$^+$, (αc) mol dm^{-3} of CH_3COO^-, and $(1 - \alpha)c$ mol dm^{-3} of CH_3COOH. Thus

$$K_c = \frac{[H^+]\,[CH_3COO^-]}{[CH_3COOH]} = \frac{(\alpha c)\,(\alpha c)}{(1 - \alpha)c} = \frac{\alpha^2 c}{1 - \alpha}$$

4.8 Effect of conditions on reaction yields and K_{eqm}

One of the major goals in carrying out chemical reactions is to maximize the yield. In a reaction that reaches equilibrium, this means that we need to know how variation of the experimental conditions affects K_{eqm}. It is therefore important for us to be able to calculate the conditions under which to carry out a reaction in order to optimize yields.

Effect of pressure

It is a common misunderstanding that K_{eqm} depends on the system pressure. This is not so. K_{eqm} depends on ΔG° which is only valid at 1 bar pressure and hence K_{eqm} *is a thermodynamic constant which does not vary at constant temperature.*

This is *not* to say that pressure has no effect on the *position* of equilibrium. Taking the example developed earlier for the $N_2O_4 \leftrightarrow 2NO_2$ system,

$$K_p = \frac{4\alpha^2 P_{tot}}{(1 - \alpha^2)}$$

If K_p is to be constant, if P_{tot} changes α must change so as to compensate. Using the approximation that for small α, $K_p \approx 4\alpha^2 P_{tot}$, it can be seen that if the total pressure is increased, the degree of dissociation of N_2O_4 will decrease by a corresponding amount so that the yield of NO_2 will go down. In a case where the pressure is varied by the addition of an inert gas that does not participate in the reaction, the pressure of the inert gas does not appear in the expression for K_p but it must be included when calculating the mole fractions of the components.

Effect of temperature

Combining eqns 4.3 and 4.24, it can be seen that

$$\ln K_{eqm} = \frac{-\Delta G^{\circ}}{RT} = \frac{-\Delta H^{\circ}}{RT} + \frac{T\Delta S^{\circ}}{RT}$$

or
$$\ln K_{eqm} = \frac{\Delta S^{\circ}}{R} - \frac{\Delta H^{\circ}}{RT} \tag{4.28}$$

This is a variation of the *Van't Hoff equation*. If we make the familiar assumption that ΔH° and ΔS° are independent of temperature over the range studied, then we can write

$$\ln K_{eqm} = (\text{constant}) - \frac{\Delta H^{\circ}}{R}\left(\frac{1}{T}\right) \tag{4.29}$$

showing that there is a linear relationship between $\ln K_{eqm}$ and $(1/T)$ as in Fig. 4.6. An alternative approach arises if the equilibrium constant is known at two temperatures T_1 and T_2,

$$\ln\left(\frac{K_{eqm}(T_2)}{K_{eqm}(T_1)}\right) = \frac{\Delta H^{\circ}}{R}\left(\frac{1}{T_1} - \frac{1}{T_2}\right) \tag{4.30}$$

Equation 4.30 allows us to correlate the effect of temperature on the reaction. If we have an exothermic reaction, so that ΔH is negative, then it is apparent that K_{eqm} will decrease as T increases. The opposite is true for an endothermic reaction.

Note: Equation 4.28 can be written as the equation of a straight line:

$$y = c + mx$$

with the slope, m, equal to $(-\Delta H^{\circ}/R)$ and the intercept equal to $\Delta S^{\circ}/R$.

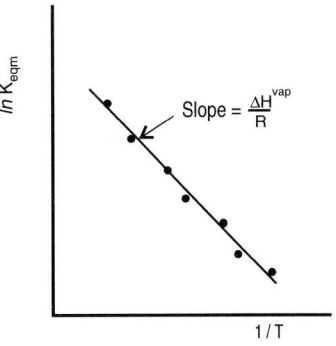

Fig. 4.6   The relationship between K_{eqm} and temperature.

Le Chatelier's principle

The above explanations provide the thermodynamic basis of a very useful principle, first suggested by Le Chatelier on the basis of empirical observations. His principle can be written as:

If a system is subjected to a constraint, it will react so as to minimize the effect of the constraint.

For example, an increase in pressure reduced the yield of NO_2 from dissociation of N_2O_4. This follows Le Chatelier's principle since the reaction moves towards the side of the equation with the lowest number of moles of gas, i.e. if the pressure is raised, the system reacts by minimizing the pressure of gas produced.

Similarly, if the temperature is increased in an exothermic reaction, the system will react so as to absorb the extra energy and so change to minimize the production of energy by the reaction. The reaction therefore proceeds to a lesser extent so that the yield is lowered. The opposite argument can be used for an endothermic reaction where the yield would be increased by raising the temperature.

Note: The systematic name for acetic acid is ethanoic acid.

Example 4.5   In the vapour phase, acetic acid partially associates into dimers. At a total pressure of 0.2 bar, acetic acid is 92% associated at 25 °C and 82% associated at 45 °C. Calculate the enthalpy and entropy changes for the reaction. What will be the effect on the dissociation of changing the total pressure?

The reaction can be represented as

$$2\text{HOAc} \leftrightarrow (\text{HOAc})_2$$

If (for convenience) we begin with 100 mol of HOAc and 92% dimerizes, then 8 mol will be present at equilibrium. The 92 mol that react give rise to 46 mol of dimer. The total number of moles present is therefore 54. Hence, using the data at 25 °C, the equilibrium constant, K_p is given by

$$K_p = \frac{p_{(\text{HOAc})_2}}{(p_{\text{HOAc}})^2} = \frac{x_{(\text{HOAc})_2} P_{\text{tot}}}{(x_{\text{HOAc}} P_{\text{tot}})^2} = \frac{(46/54)\,0.2}{[(8/54)\,0.2]^2} = 194.1$$

Using the data at 45 °C gives $K_p = 37.33$. From eqn 4.22, $\Delta G^\circ = -13.06$ kJ mol^{-1} at 25 °C and -9.576 kJ mol^{-1} at 45 °C.

We can now apply eqn 4.30

$$\ln\left(\frac{K_{\text{eq}}(T_2)}{K_{\text{eq}}(T_1)}\right) = \frac{\Delta H^\circ}{R}\left(\frac{1}{T_1} - \frac{1}{T_2}\right);$$

$$\ln\left(\frac{3.446}{1.021}\right) = \frac{\Delta H^\circ}{8.314}\left(\frac{1}{298.15} - \frac{1}{318.15}\right)$$

Exercise: Express the equation for K_p in terms of the degree of association, α, and the total pressure, and so prove that α decreases with increasing pressure.

Hence, $\Delta H^\circ = -65.0$ kJ mol^{-1}. Applying eqn 4.3 at 45 °C gives $\Delta S^\circ = -174.2$ J K^{-1} mol^{-1}.

Applying Le Chatelier's principle, an increase of pressure should favour the side of the reaction with the lower number of gas moles, i.e. favour the dimerization.

Note that, as expected, the extent of dimerization decreases as the temperature rises since it is an endothermic reaction. The dimerization must lead to an increase in order, hence the negative entropy change. An alternative method for the calculation is to write the two equations

$$\Delta G^\circ(25^\circ C) = -13.06 \text{ kJ mol}^{-1} = \Delta H^\circ - (298.15)\Delta S^\circ$$

$$\Delta G^\circ(45^\circ C) = -9.58 \text{ k J mol}^{-1} = \Delta H^\circ - (318.15)\Delta S^\circ$$

and solve the equations simultaneously assuming that ΔH° and ΔS° do not vary with temperature.

4.9 Kinetics and thermodynamics of equilibrium

The discussion so far in this primer has concentrated on thermodynamic aspects of equilibrium. There is a different approach, perhaps more intuitively straightforward, which involves kinetic arguments. An alternative definition of equilibrium is the composition at which the forward and reverse reactions proceed at the same rate. This also emphasizes the *dynamic* nature of chemical equilibrium. The reaction does not stop when equilibrium is reached but the overall proportion of reactants and products does not change.

If we have the general reaction

$$A + B \leftrightarrow C$$

then the rate of the forward reaction is given by k_f [A] [B] and that of the reverse reaction by k_b [C]. At equilibrium, these are equal so that:

$$k_f[A]_{eqm}[B]_{eqm} = k_b[C]_{eqm} \text{ or } \frac{k_f}{k_b} = \frac{[C]_{eqm}}{[A]_{eqm}[B]_{eqm}} = K_c \qquad (4.31)$$

Thus, the ratio of the forward and reverse rate constants also gives the equilibrium constant and provides a link between kinetic and thermodynamic arguments applied to chemical reactions. Note that this has been applied to the derivation of K_c. An analogous argument using partial pressures in place of concentration and the appropriate thermodynamic standard states leads to an expression of the same form for the relation between K_p and the forward and reverse rate constants for a gas reaction.

It is also appropriate to note that kinetic arguments must also be considered in determining the optimum conditions for a particular process. For example, with an exothermic reaction, Le Chatelier's principle, as discussed above, suggests that the yield would be maximized by reducing the temperature. However, reactions are generally slower at lower temperatures. A process chemist must balance the use of reagents and the yield of products with the length of time for which it must be carried out and the energy demand, in order to optimize a reaction.

Note: The derivation of rate equations is dealt with in detail in Primer No. 21 by Cox. For now it is sufficient to note that the rate of reactions is given by the product of the rate constant, k, and the concentration of reactant.

4.10 Thermodynamics of electrochemical cells

An important class of reaction not so far included is that of electron transfer or *redox* reactions. A complete discussion would be out of place here and has

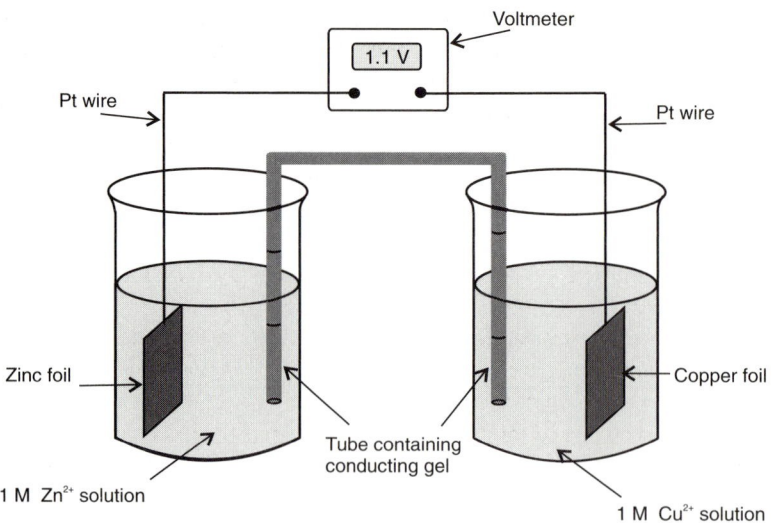

Fig. 4.7   An electrochemical cell for the reaction

$$Zn_{(s)} + Cu^{2+}_{(aq)} \rightarrow Cu_{(s)} + Zn^{2+}_{(aq)}$$

appeared in other primers in the series, but it is appropriate to focus on thermodynamic aspects of these reactions.

For example, if a piece of pure zinc metal is placed in an aqueous solution of copper sulfate, the zinc rapidly becomes coated with a brown deposit of metallic copper. The chemical reaction taking place is

$$Zn_{(s)} + Cu^{2+}_{(aq)} \rightarrow Cu_{(s)} + Zn^{2+}_{(aq)}$$

In fact, this can be split into two *half-reactions*

$$Zn_{(s)} - 2\,e \rightarrow Zn^{2+}_{(aq)} \quad \textit{oxidation}$$
$$Cu^{2+}_{(aq)} + 2\,e \rightarrow Cu_{(s)} \quad \textit{reduction}$$

Reactions involving electron gain are reductions while those involving electron loss are oxidations. Clearly, there must be a net balance of electrons and so both processes must take place concurrently, hence the name *redox* reaction.

If the two reactions are carried out in separate containers (although linked to allow electron or charge transport) electrons can be made to flow around an external circuit. Alternatively, if the circuit is set up as in Fig. 4.7, so that it includes a high resistance such as a voltmeter, electron flow is prevented and the potential difference between the electrodes which provides the 'driving force' for the electron flow can be measured. Under these conditions, the cell acts reversibly.

Standard electrode potentials

Note: Further details of electrode potentials and the SHE can be found in Primer No. 41 by Compton and Sanders

The voltage generated clearly depends on the nature of the two half-cells. Again, to assign an absolute value of the potential of a half-cell, we need to fix a standard. This is done by setting to zero the potential for a half-cell involving the reaction

$$H^{+}_{(aq)} + e \rightarrow \tfrac{1}{2} H_{2(g)}$$

at 298 K, 1 bar of hydrogen pressure, and 1 mol dm^{-3} concentration of H$^+$. Construction of a cell in which one half consists of this *standard hydrogen electrode*, *SHE*, allows the potential for any other half-cell to be determined. Extensive lists of these values have been published and can be found in the references given in Further Reading. Electrode potentials measured under these standard conditions are given the symbol $E°$, while E represents the voltage generated when the cell operates under conditions other than those specified above.

The emf of the cell, E, tells us about the tendency for electrons to flow around the circuit, which is clearly the same as the tendency for the redox reaction to occur. We saw earlier that the tendency for a reaction to happen is correlated to the free energy change, ΔG. There must then be a direct connection between E and ΔG.

If n electrons are transferred in the reaction, then the total charge, Q, is nF. F is the Faraday constant which is the total charge on 1 mole of electrons, numerically equal to 96485 °C mol^{-1}. We introduced briefly, in Chapter 1, that the electrical work done when a charge Q moves through a potential difference E is given by

$$w_{\text{elec}} = \int_{Q_{\text{initial}}}^{Q_{\text{final}}} -E \, \mathrm{d}Q \qquad (1.7)$$

If the potential difference, or voltage, is constant, the work done in transferring a charge is simply

$$w_{\text{elec}} = -nFE$$

At constant temperature and pressure, this work is done as a result of the lowering of the system free energy as it moves towards equilibrium. Thus, we can write

$$\Delta G = nFE \qquad (4.32)$$

In the particular case of standard conditions and all components in their standard states, then

$$\Delta G° = -nFE° \qquad (4.33)$$

Since the cell potential, E, and the free energy change, ΔG, have opposite sign, we can see that if a cell has a positive potential, the reaction taking place will be spontaneous. One of the most useful features of eqn 4.32 is that it allows us to make thermodynamic measurements on systems on which it would be difficult to perform calorimetry (see Example 4.6). Particular examples include biochemical systems, which involve a large number of redox reactions.

From eqn 4.8, we can see that

$$\left(\frac{\mathrm{d}G}{\mathrm{d}T}\right)_p = -S \quad \text{or} \quad \Delta S° = -\left(\frac{\mathrm{d}\Delta G°}{\mathrm{d}T}\right)_p \qquad (4.34)$$

Substituting eqn 4.33,

$$\Delta S° = nF\left(\frac{\mathrm{d}E°}{\mathrm{d}T}\right)_p \qquad (4.35)$$

Note: Again the definition of 1 mol dm^{-3} as the standard concentration is an approximation and it should be defined as *unit activity*. However, the treatment of activities in aqueous solution is outside the scope of this primer and we will retain this simplification.

so that, as long as we are operating at constant pressure, measurement of the temperature variation of the cell voltage allows us to measure the entropy change for the reaction. Once ΔG° and ΔS° are known from eqn 4.32 and 4.35, ΔH° can be calculated.

Example 4.6   An electrochemical cell with the following reaction generates a standard electrode potential of 1.055 V at 25 °C and 1.015 V at 0 °C. Calculate the standard enthalpy, entropy, and free energy changes for the reaction at 25 °C.

$$Zn_{(s)} + 2AgCl_{(s)} \rightarrow ZnCl_{2(aq)} + 2Ag_{(s)}$$

The standard free energy change at 298 K follows directly from eqn 4.32.

$$\Delta G^\circ = -nFE^\circ = -(2)(96485)(1.055) = -203.58 \text{ kJ mol}^{-1}$$

We will make the approximation that the voltage is a linear function of temperature. The standard entropy change arises from the temperature dependence via eqn 4.35

$$\Delta S^\circ = nF\left(\frac{dE^\circ}{dT}\right)_p = (2)\,(96485)\left(\frac{1.055 - 1.015}{298.15 - 273.15}\right) = 308.75 \text{ J K}^{-1} \text{ mol}^{-1}$$

ΔH° can then be found

$$\Delta H^\circ = \Delta G^\circ + T\Delta S^\circ = (-203.58 \times 10^3) + (298.15)(308.75)$$
$$= -111.53 \text{ kJ mol}^{-1}$$

Concentration dependence of electrode potentials

We saw earlier that the free energy dependence of reactions in solution can be described by eqn 4.21

$$\Delta G = \Delta G^\circ + RT \ln Q$$

Substituting eqn 4.32 and eqn 4.33,

$$nFE = -nFE^\circ + RT \ln Q \quad \text{or} \quad E = E^\circ - \frac{RT}{nF} \ln Q \qquad (4.36)$$

This is known as the *Nernst equation*. When the cell reaction reaches equilibrium, $Q = K_{eqm}$ and electron flow stops and $E = 0$. Thus, it follows that

$$E^\circ = \frac{RT}{nF} \ln K_{eqm} \qquad (4.37)$$

This provides a very convenient method for estimating the equilibrium constants and standard free energy changes, since E° values are available for a very wide range of reactions.

4.11   Free energy changes in biochemical systems

The ideas on free energy changes and equilibria that have been discussed in this chapter, can be applied to all types of reaction. However, those occurring in living cells occur under rather special conditions which are worthy of further comment.

Biochemists' standard state

The conditions used to define the standard state for chemical reactions in solution were defined in Section 4.3 as a temperature of 25 °C and a concentration of 1 mol dm^{-3}. These conditions are not convenient for many biochemical reactions. For example, a large number of biochemical processes involve the transfer of H$^+$ ions (*proton transfers*) and the usual standard concentration would correspond to pH = 0. This would be very unusual in living systems where most cells have a *physiological pH* close to neutral, pH = 7.

This leads us to define an alternative *biochemical standard state* where the standard state for [H$^+$] is taken as 1×10^{-7} mol dm^{-3}. Standard free energy changes occurring under these conditions are designated $\Delta G^{\circ\prime}$.

Consider a reaction of the form

$$A + B \rightarrow C + n\,H^+$$

These occur in aqueous solution for biochemical systems so that the (aq) subscripts will be omitted for this section. The Gibbs free energy change for the reaction is given by

$$\Delta G = \Delta G^\circ + RT \ln\left(\frac{([C]/1M)\,([H^+]/1M)^n}{([A]/1M)\,([B]/1M)}\right)$$

where 1 M represents 1 mol dm^{-3}. If we now use the biochemists' standard state, this must be rewritten as

$$\Delta G = \Delta G^{\circ\prime} + RT \ln\left(\frac{([C]/1M)\,([H^+]/10^{-7}M)^n}{([A]/1M)\,([B]/1M)}\right)$$

so that it is clear that

$$\Delta G^\circ = \Delta G^{\circ\prime} + nRT \ln(1/10^{-7}) \tag{4.38}$$

At 298 K and for $n = 1$, $\Delta G^\circ = \Delta G^{\circ\prime} + 40.0$ kJ mol^{-1}. Thus, for this type of reaction ΔG° is greater than $\Delta G^{\circ\prime}$ by 40 kJ mol^{-1} for each mole of H$^+$ involved. Hence, the reaction equilibrium favours the products more at pH 7 than pH 0. The reverse is clearly true for reactions involving the consumption of H$^+$.

In addition to the above considerations on the standard concentrations in biochemical systems, it should be remembered that most mammals have body temperatures above 25 °C, human biochemistry taking place around 37 °C. Thermodynamic data is therefore often quoted at 37 °C although, where needed, correction of data from 298 K is straightforward using the methods introduced in Section 4.3.

Coupled reactions

The discussion throughout this chapter has centred on the use of Gibbs free energy changes to predict the course of reactions. However, thus far, single reactions have been considered in isolation. A reaction with positive ΔG will not be spontaneous. It can be made to occur though if it is *coupled* to another reaction with a larger, negative ΔG. Such reactions are sometimes called *tandem* reactions.

The most important example of this class of reactions arises from metabolic processes. The 'driving force' for metabolism arises from the two molecules

Note: The pH of a solution is defined by pH = $-\log_{10}$ ([H$^+$]/mol dm^{-3}). Hence,

pH $= 0 \equiv$ [H$^+$] $= 1$ mol dm^{-3}
pH $= 1 \equiv$ [H$^+$] $= 10^{-1}$ mol dm^{-3}
pH $= 7 \equiv$ [H$^+$] $= 10^{-7}$ mol dm^{-3}
pH $= 14 \equiv$ [H$^+$] $= 10^{-14}$ mol dm^{-3}

Mathematical note: $10^0 = 1$.

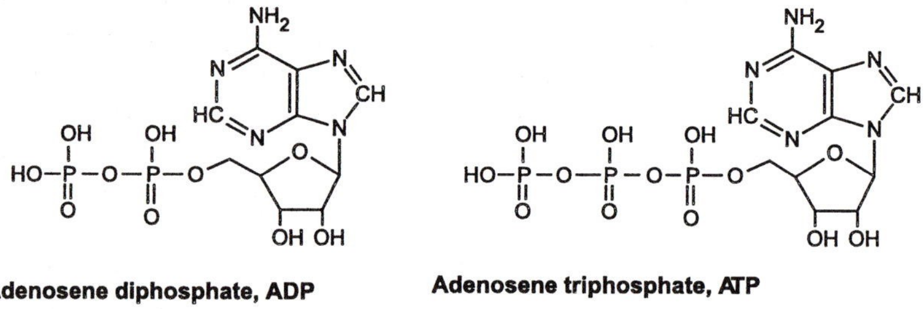

Fig. 4.8 The chemical structure of adenosine triphosphate, ATP. ADP has one less phosphate in the side-chain.

adenosine diphosphate, ADP, and its triphosphate analogue, ATP, illustrated in Fig. 4.8. The terminal phosphate group can be readily hydrolysed

$$\text{ATP}_{(aq)} + \text{H}_2\text{O}_{(l)} \rightarrow \text{ADP}_{(aq)} + \text{H}^+_{(aq)} + \text{P}^-_{(aq)} \qquad (1)$$

where P^- indicates an inorganic phosphate group (including PO_4^{3-}, H_2PO_4^-, etc.) For this reaction, $\Delta G^{\circ\prime} = -30.5$ kJ mol^{-1} at 37 °C. This relatively large change in free energy is therefore available to 'drive' other processes.

A major source of energy in living cells arises from the oxidation of sugars. The first step in the metabolism of glucose, which ends up being converted to lactic acid when performed anaerobically (i.e. in the absence of sufficient oxygen for complete reaction) is the reaction of a free hydroxyl group on the sugar with a phosphate ion to form glucose-6-phosphate.

$$\text{glucose}_{(aq)} + \text{P}^-_{(aq)} \rightarrow \text{glucose-6-phosphate}_{(aq)} + \text{H}_2\text{O}_{(l)} \qquad (2)$$

$\Delta G^{\circ\prime}$ for this reaction step is $+13.4$ kJ mol^{-1} at 37 °C so that it is not spontaneous. Remember that, strictly, this value refers to all components of the reaction in their standard state concentrations. However, it is easy to show that the Gibbs free energy change calculated at concentrations appropriate to cells will not change the sign of $\Delta G^{\prime}$. The principle by which reaction (2) can occur is that it is *coupled* to reaction (1). Adding these together gives

$$\text{glucose}_{(aq)} + \text{ATP}_{(aq)} \rightarrow \text{glucose-6-phosphate}_{(aq)} + \text{ADP}_{(aq)} + \text{H}^+_{(aq)} \qquad (3)$$

for which the overall $\Delta G^{\circ\prime}$ is $(-30.5 + 13.4) = -17.1$ kJ mol^{-1}, so the overall reaction is spontaneous. The regeneration of ATP from ADP relies on another series of coupled reactions.

The really impressive trick that living cells play is to ensure that reactions (1) and (2) are coupled together in this form rather than the phosphate ions reacting with any other species in solution. This is the role of the *enzymes*, which are protein catalysts that are extremely specific for each reaction. In this case the *hexokinase* enzyme binds the glucose and a phosphate ion in such a way that they react together rather than being free to react with other cell components.

4.12 Concluding remarks

The ideas introduced in this chapter are some of the most fundamental and useful throughout physical chemistry. Hopefully, you will already appreciate some of the things that the Gibbs free energy can tell us about a chemical

Note: The complete conversion of glucose to lactic acid is a complex pathway involving nine intermediate compounds. A number of these are driven by the reaction of ATP to ADP. In others, the reverse reaction takes place, regenerating some ATP molecules. When a cell is supplied with sufficient oxygen to perform the oxidation *aerobically*, the sequence is more complex and the end result is complete oxidation to carbon dioxide and water according to

$$\text{C}_6\text{H}_{12}\text{O}_6 + 38\,H^+ + 38\,ADP + 38\,P \rightarrow$$
$$38\,\text{ATP} + 6\text{CO}_2 + 44\text{H}_2\text{O}$$

Note that here the overall reaction *produces* ATP.

system. You will meet many others in further study of chemistry. The major use here has been to predict the equilibrium state of chemical reactions, from which some idea of reaction yields can be gained. However, it is important to realize that ΔG is not the whole story. Many chemical situations that arise are far from equilibrium and other methods (so-called *irreversible* or *time-dependent* thermodynamics—well outside the scope of this primer) must be used. In addition, kinetic factors must always be considered.

This chapter has focused on applying the ideas of Gibbs free energy to chemical reactions. In the next chapter, its use will be further illustrated in studying the phase behaviour of compounds. To complete this chapter, a further example is presented which brings together the concepts of Gibbs free energy, equilibrium, and the estimation of product yields from a chemical reaction.

Example 4.7 The standard enthalpy of formation of water is $-241.8 \text{ kJ mol}^{-1}$. Use the mean heat capacities and standard entropies listed in Appendix 1 to estimate the percentage dissociation of water vapour at 2000 °C and 0.01 bar pressure.

The reaction of interest is the *dissociation* of water

$$H_2O_{(g)} \rightarrow H_{2(g)} + \tfrac{1}{2}O_{2(g)}$$

At 25 °C, $\Delta H_{298}^{\circ} = -\Delta H_{f,298}^{\circ} = +241.8 \text{ kJ mol}^{-1}$

$$\Delta S_{298}^{\circ} = [130.7 + \tfrac{1}{2}(205.1)] - [188.3] = 44.95 \text{ J mol}^{-1} \text{ K}^{-1}.$$

To convert ΔH_{298}° and ΔS_{298}° to the higher temperature, we need $\Delta \bar{c}_{p}$.

$$\Delta \bar{c}_{p} = [28.8 + \tfrac{1}{2}(29.4)] - [33.6] = 9.9 \text{ J mol}^{-1} \text{ K}^{-1}.$$

At 2000 °C, 2273 K.

$$\Delta H_{2273}^{\circ} = \Delta H_{298}^{\circ} + \Delta \bar{c}_{p}(2273 - 298) = 241.8 + (9.9 \times 10^{-3})(1975)$$
$$= 261.35 \text{ kJ mol}^{-1}.$$
$$\Delta S_{2273}^{\circ} = \Delta S_{298}^{\circ} + \Delta \bar{c}_{p} \ln(2273/298) = 44.4 + (9.9) \ln(7.628)$$
$$= 65.06 \text{ J mol}^{-1} \text{ K}^{-1}.$$
$$\Delta G_{2273}^{\circ} = \Delta H_{2273}^{\circ} - T\Delta S_{2273}^{\circ} = 261.35 - (2273)(65.06 \times 10^{-3}) \text{ kJ mol}^{-1}.$$
$$= +113.46 \text{ kJ mol}^{-1}.$$
$$K_p = \exp(-\Delta G^{\circ}/RT) = \exp(-113.46 \times 10^3/2273\,R) = 2.47 \times 10^{-3} \text{ bar}$$

If we start with 1 mol of water and the degree of dissociation is α, then at equilibrium there will be α mol of H_2, $\tfrac{1}{2}\alpha$ mol of O_2, and $(1 - \alpha)$ mol of H_2O.

$$K_p = \frac{(p_{H_2})(p_{O_2})^{\frac{1}{2}}}{(p_{H_2O})} = \frac{(x_{H_2}p_{tot})(x_{O_2}p_{tot})^{\frac{1}{2}}}{(x_{H_2O}p_{tot})}$$

$$= \left(\frac{(\alpha)}{(1+0.5\alpha)}\right)\left(\frac{(0.5\alpha)^{\frac{1}{2}}}{(1+0.5\alpha)^{\frac{1}{2}}}\right)\left(\frac{(1+0.5\alpha)}{(1-\alpha)}\right)\frac{p^{\frac{3}{2}}}{p} = \frac{\sqrt{0.5}\,\alpha^{\frac{3}{2}}p^{\frac{1}{2}}}{(1-\alpha)(1+0.5\alpha)^{\frac{1}{2}}}$$

If we now make the assumption that α is small, then $\alpha \ll 1$, $(1 - \alpha) \approx (1 + 0.5\,\alpha) = 1$ so that

$$K_\text{p} = \frac{\sqrt{0.5}\,\alpha^{\frac{3}{2}} p^{\frac{1}{2}}}{1} = 0.7071\,(0.1)\,\alpha^{\frac{3}{2}} = 2.47 \times 10^3$$

Hence $\alpha \approx 0.00653$ or 0.653%.

Note that, as well as the mathematical approximation in the final step, the result is an approximation in that the mean heat capacities used are unlikely to give an accurate result over such a large temperature range. However, for general use such an approach can be useful.

5 Phase equilibrium and solutions

The concept of matter existing in one of three phases—*solid*, *liquid*, or *gas*—is familiar to us and we have already discussed some of the thermodynamic properties, such as the changes of enthalpy and entropy, that take place when phase changes occur. To begin, we must be clear in thermodynamic terms what we mean by a phase. A precise definition is that a *phase is part of a system which is homogeneous throughout and separated from other phases by a definite boundary*. For gases or vapours, there can be only one phase irrespective of the number of components involved. For a single component there can be only one liquid phase. However, if there are two or more liquids, the phase behaviour is more complex. The liquids can mix completely in all proportions, e.g. ethanol and water. Alternatively, they may not mix at all, e.g. oil and water which form two phases, the surface between them being the phase boundary. Other mixtures may mix under some circumstances but not others. Phase behaviour is more complex in solids, where even a single component may exhibit a large number of phases. Familiar examples are carbon, which can exist as graphite, diamond, or fullerene, and sulphur, which can exist as small rings or as one of several crystal allotropes.

The aim of this chapter is to investigate the factors which determine why compounds exist in a particular phase under a particular set of conditions and which properties influence the transitions between phases.

5.1 Phase behaviour of single components

As developed in the previous chapter, a system will always tend to move towards a state of minimum Gibbs free energy. In broad terms, therefore, the phase that exists will be that with the lowest G under the ambient conditions.

For a single, pure component, the dependence of Gibbs free energy on temperature is given by eqn 4.8 and has the form shown schematically in Fig. 5.1. The slopes of the lines give the entropy of each phase. Solids have lower entropy than liquids which have lower values than gases, and this is reflected in the plots.

At the temperatures where the lines intersect, transitions between phases occur. If, as an example, we focus on melting, at the melting point, T_m, the Gibbs free energies of the solid and liquid must be equal. Therefore,

$$\Delta G^{fus} = G^{liquid} - G^{solid} = 0 = \Delta H^{fus} - T_m \Delta S^{fus}$$

or

$$\Delta S^{fus} = \frac{\Delta H^{fus}}{T_m} \tag{5.1}$$

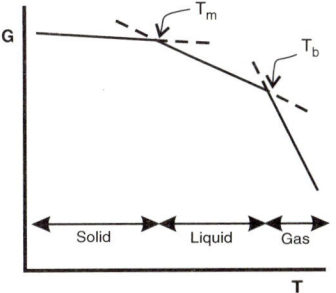

Fig. 5.1 Gibbs free energy as a function of temperature for a pure component.

The same argument can be applied for the vaporization at T_b, so that

$$\Delta S^{\text{vap}} = \frac{\Delta H^{\text{vap}}}{T_b} \tag{5.2}$$

Single component phase diagrams

The major experimental variables which affect phase behaviour are the pressure and temperature. At sufficiently low temperatures and/or high pressures, matter exists in the solid phase. At high temperatures and low pressures, the stable phase is the gas. *Phase diagrams* give a convenient way of representing the phase behaviour. The easiest way to appreciate what phase diagrams can tell us is to consider how they are constructed experimentally.

This is done by measuring, at a large number of different pressures, the melting and boiling points (or, if appropriate, sublimation temperatures) of a component. These can then be plotted as a function of pressure as indicated schematically in Fig. 5.2. For example, if we begin with a component in the gas phase at pressure p_1 and allow it to cool, it will condense at the boiling point and then freeze when it reaches the melting (or freezing) point. This gives two points on the phase diagram. The measurements of T_m and T_b can be repeated over a range of pressures (e.g. p_2 and p_3 to construct the remainder of the diagram. At sufficiently low pressures (e.g. p_4) the intermediate liquid may not be formed and the compound will condense directly to the solid.

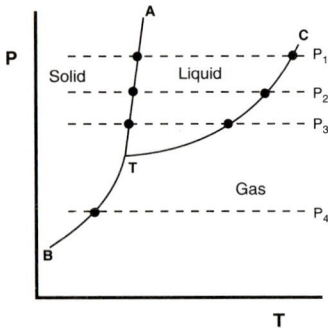

Fig. 5.2 Schematic construction of a single component phase diagram.

Phase diagrams convey a large amount of information about the phase behaviour of a component. The lines indicate the sets of conditions where two phases are in equilibrium. Of more practical use, they also show how the phase transitions vary. For example, line A–T shows how the melting point changes with pressure. Line T–C shows the variation of boiling point with pressure, or, as an alternative interpretation, how the vapour pressure of a liquid depends on temperature.

The point at T, where all three phases are in equilibrium, is the *triple point*. This represents a unique set of conditions and is characteristic of a pure substance. For example, the values for water are 273.16 K and 4.58 Torr. Under no other conditions can ice, water, and water vapour exist in equilibrium. The constancy of triple points has led to their use as standards for calibrating thermometers. In addition, the triple point represents the lowest temperature at which the liquid phase of a compound can exist at any pressure.

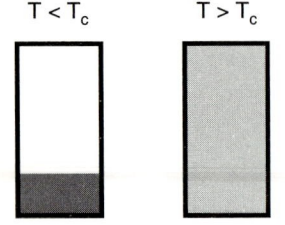

Fig. 5.3 Schematic representation of fluid densities above and below the critical conditions.

Point C is the *critical point*. This represents the maximum temperature at which a gas can be liquefied merely by increasing the pressure. If a liquid is heated in a sealed container, it will not boil. Rather, the density of the vapour increases as more liquid evaporates until, at the critical point, it becomes equal to that of the liquid. At this point, all distinction between liquid and vapour disappears, as shown schematically in Fig. 5.3. Fluids with temperatures and pressures above the critical values are known as *supercritical fluids*, SCFs. These possess some of the properties of a liquid, such as a relatively high density and ability to act as a solvent. In contrast, some properties, such as the viscosity and diffusion coefficient, are closer to those of a gas. Although regarded as merely a curiosity for many years, SCFs have become commercially useful in recent years, with carbon dioxide being a particularly useful example because of the relatively mild conditions under which it forms a SCF (see Table 5.1). Commercial plants have been constructed to use SCF-CO_2 to extract flavourings and other food components (e.g. caffeine from coffee beans, where it has largely replaced the chlorinated solvents that used to be employed). Recent research has sought to use it as a solvent for chemical reactions, taking advantage of its lack of toxicity and the ease of recovery of the products—the CO_2 just evaporates as the pressure is reduced to atmospheric.

Phase diagrams for several compounds are shown in Fig. 5.4 and several comments can be made to illustrate their utility.

- The triple point for carbon dioxide is 5.2 atm. Thus, at atmospheric pressure (0.987 bar) liquid CO_2 cannot exist. This explains the common observation that solid CO_2 undergoes sublimation in the laboratory.
- The diagram for sulphur has an additional line representing the transition between the two crystalline phases that it displays. The lines in this case show the temperatures and pressures at which transition between these phases occur. A similar phenomenon can be seen for carbon. The phase diagram for water shows that, at very high pressures, a number of different forms of ice which have different crystal structures can exist.

Table 5.1 Selected values of critical temperatures and pressures

	T_c/K	P_c/bar
Argon	150.7	48.0
Methane	190.6	45.6
Hydrogen	33.23	12.8
Water	647.4	218.3
CO_2	304.2	72.9
Xenon	289.8	58.0
Oxygen	154.8	50.1
Ammonia	405.5	111.3

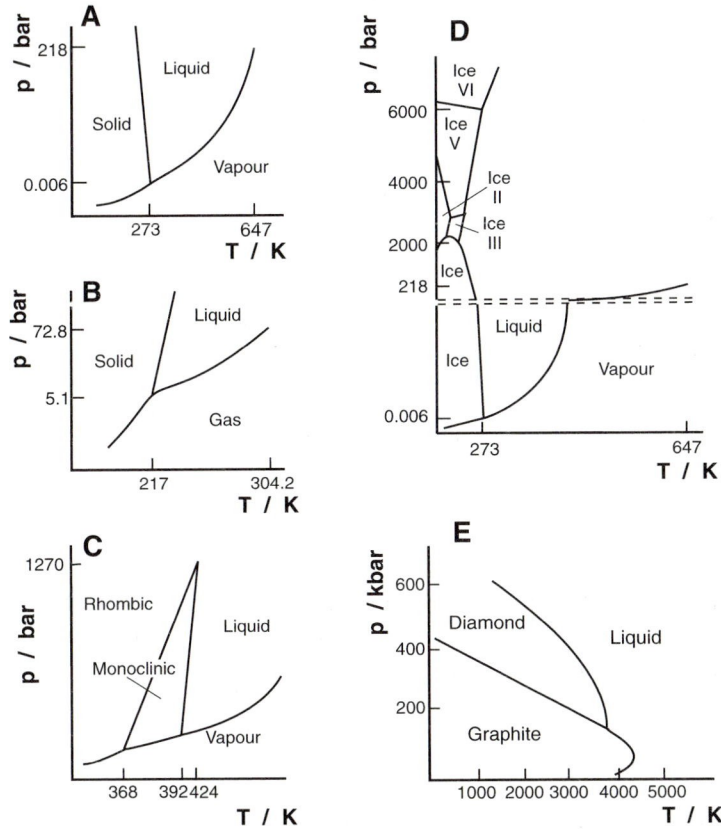

Fig. 5.4 Phase diagrams for pure components.
A: Water (at relatively low pressures). B: Carbon dioxide. C: Sulphur
D: Water (for a wide range of pressures — note the change of scale). E: Carbon.

- It is notable that the slopes of the solid–liquid lines are steeper than those of the liquid–vapour lines. The latter is also distinctly more curved.
- One noticeable feature of the diagram for water is that the line for the solid–liquid transition has an opposite slope to that in the other cases.

These observations are merely noted here and will be accounted for in the following sections.

5.2 Quantitative treatment of phase transitions

First, let us think about why phases exist in equilibrium. Consider a single component at equilibrium distributed between two phases, phase I and phase II, under conditions of temperature T and pressure p, as in Fig. 5.5. If a small amount, dn mol, of material is transferred from phase I to phase II, then from the discussion in Chapter 4, the changes in Gibbs free energy, dG, are given by

$$dG^{I} = -\mu^{I}dn \quad dG^{II} = \mu^{II}dn$$

so that the total change in Gibbs free energy is $dG = (\mu^{II} - \mu^{I})dn$.

When the system reaches equilibrium, $dG = 0$ and so it follows that the chemical potential is the same in each phase, i.e. $\mu^{II} = \mu^{I}$. The argument can be generalized to any number of phases so that we can define, as a condition for phase equilibrium,

> *At constant* p *and* T, *a system comes to equilibrium when the chemical potential of a component is the same in each phase.*

Effect of temperature and pressure

We usually think about phase transitions in terms of temperature changes, e.g. heating ice to form water, or by reducing the pressure, e.g. vacuum evaporation of solvents. These are the situations described by the phase diagrams in the previous section and we now need a quantitative model to describe these changes.

If we again consider the situation described in Fig. 5.5, with two phases in equilibrium, and change the conditions to $(T + dT)$ and $(P + dP)$. Then the change in Gibbs free energy, dG, can be described for each phase by eqn 4.7.

$$dG^{I} = V^{I}dP - S^{I}dT \quad \text{and} \quad dG^{II} = V^{II}dP - S^{II}dT$$

If the phases are to remain in equilibrium under the new conditions, the change in free energy must be the same in each phase, $dG^{I} = dG^{II}$. Hence

$$V^{I}dP - S^{I}dT = V^{II}dP - S^{II}dT$$
$$(V^{I} - V^{II})dP = (S^{I} - S^{II})dT \quad \text{or} \quad \Delta V dP = \Delta S dT$$

where ΔV and ΔS are, respectively, the molar changes in volume and entropy that occur during the transition. We can rearrange this equation and substitute for ΔS using eqn 5.1 to show

$$\frac{dP}{dT} = \frac{\Delta S}{\Delta V} = \frac{\Delta H}{T \Delta V} \tag{5.3}$$

Equation 5.3 is the *Clapeyron equation* and applies to both the melting and vaporization transitions in single components. In particular, for the former, we can write it in the form

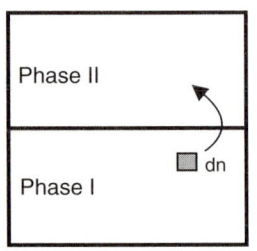

Fig. 5.5 Transfer of a component between phases at equilibrium.

Note: Although we will not prove it here, this condition can be generalized for any number of components in any number of phases. It can formally be stated that:

The condition for phase equilibrium is that the chemical potential of each component is the same in all phases.

$$\frac{dP}{dT} = \frac{\Delta H^{\text{fus}}}{T_{\text{m}} \Delta V^{\text{fus}}} \qquad (5.4)$$

where ΔV^{fus} is the change in volume on melting. The volume change can be calculated from the densities of the solid and liquid at the melting point. In the form of a differential equation, as eqn 5.3, the Clapeyron equation is exact. In practice, we make the reasonable assumption that ΔH^{fus} does not vary with temperature so that (dP/dT) is constant and P is a linear function of T. This explains the form of the solid–liquid lines in the phase diagrams in the previous section. Additionally, it allows us to determine (dP/dT) from experimental measurements (see Example 5.1).

Since ΔH^{fus} and T_{m} must have positive values, the sign of ΔV^{fus} determines the sign of (dP/dT) and hence the slope of the solid–liquid line on the phase diagram. The vast majority of substances have higher densities in the solid phase than the liquid so that ΔV^{fus}, and hence the slope, is positive. This indicates that the melting point rises with increasing pressure. The major exception to this is water where the collapse of the relatively open, low-density structure of ice causes a contraction in volume on melting to water and a consequent increase in density on melting. Thus, for water, the behaviour is the opposite to most compounds and the melting point is reduced as the pressure increases. This explains the observation that the solid–liquid line in the phase diagram slopes in the opposite direction from that for most compounds.

One point worthy of note is that for most compounds, the density difference between solid and liquid, and hence ΔV^{fus}, is small so that (dP/dT) is large. This means that large changes of pressure, typically of the order of several tens of atmospheres, are needed to significantly change T_{m}.

Note: The density, ρ, of a substance is defined as its mass divided by its volume and is usually reported in g cm^{-3} or kg m^{-3}. Thus, the *specific* volume V_{s} is given by $(1/\rho)$ and has units of cm^3 g^{-1}. The *molar* volume, V° measured in cm^3 mol^{-1} is given by (molar mass/density) = (M/ρ).

Example 5.1 The melting point of sodium (molar mass 22.99 g mol^{-1}) is 97.8 °C at 1 atm pressure. The densities at this temperature of the solid and liquid are 0.929 g cm^{-3} and 0.952 g cm^{-3} respectively. The enthalpy of fusion is 3 kJ mol^{-1}. Calculate the melting point of sodium at a pressure of 120 atm.

Using the Clapeyron equation, eqn 5.3,

$$\frac{dP}{dT} = \frac{\Delta H^{\text{fus}}}{T_{\text{m}} \Delta V^{\text{fus}}} = \frac{3 \times 10^3 \text{ J mol}^{-1}}{(370.95 \text{ K}) \Delta V^{\text{fus}}}$$

ΔV^{fus} is the difference in molar volumes for liquid and solid.

$$\Delta V^{\text{fus}} = V^{\circ}(\text{liquid}) - V^{\circ}(\text{solid}) = \left(\frac{22.99 \text{ g mol}^{-1}}{0.929 \text{ g cm}^{-3}} - \frac{22.99 \text{ g mol}^{-1}}{0.952 \text{ g cm}^{-3}} \right)$$

$$= 0.598 \text{ cm}^3 \text{ mol}^{-1}$$

$$\frac{dP}{dT} = \frac{3 \times 10^3 \text{ J mol}^{-1}}{(370.95 \text{ K})(0.598 \times 10^{-6} \text{ m}^3)} = 1.352 \times 10^7 \text{ Pa K}^{-1}$$

We now assume that $\dfrac{dP}{dT}$ is given by the macroscopic changes $\dfrac{\Delta P}{\Delta T}$. Hence

$$\frac{\Delta P}{\Delta T} = 1.352 \times 10^7 \text{ Pa K}^{-1} \quad \text{or} \quad \Delta T = \frac{(120 - 1)\text{atm} \times (101325 \text{ Pa atm}^{-1})}{1.352 \times 10^7 \text{ Pa K}^{-1}}$$

$\dfrac{\Delta T}{^\circ C} = 0.892$ K. Thus, T_m changes by 0.892 K so that T_m at 120 atm is 98.7 °C.

The reason for the interest in liquid sodium is that it is used as a coolant in some types of power-generating nuclear reactors. Clearly, solidification under these conditions would have undesirable consequences.

While the Clapeyron equation is, in principle, applicable to all phase changes in single component systems, we can recast the equation into a more useful form when one of the phases is a gas. The major difference for vaporization or sublimation is that the change in volume will be much larger. Therefore, we would expect pressure changes to have a larger influence on T_b or T_{sub}.

We introduce three assumptions into eqn 5.4. First, as in previous work, we assume that the volume per mole of the solid or liquid is insignificant compared with that of the gas. Thus, $\Delta V^{vap} = V^\circ(\text{gas}) - V^\circ(\text{liquid}) \approx V^\circ(\text{gas})$. Secondly, we assume that the gas behaves ideally so that $V^\circ(\text{gas}) = RT/P$ for 1 mole. Finally, we assume that ΔH^{vap} does not vary significantly over the temperature range of interest and can therefore be treated as a constant.

The Clapeyron equation can therefore be modified to

$$\frac{dP}{dT} = \frac{\Delta H^{vap}}{T_b \Delta V^{vap}} = \frac{\Delta H^{vap}}{T_b \left(\dfrac{RT}{P}\right)} = \frac{P\,\Delta H^{vap}}{R\,T^2}$$

$$\frac{dP}{P} = \left(\frac{\Delta H^{vap}}{R}\right)\frac{dT}{T^2}$$

If the gas pressures are P_1 at temperature T_1 and P_2 at T_2, then

$$\int_{P_1}^{P_2} \frac{dP}{P} = \left(\frac{\Delta H^{vap}}{R}\right)\int_{T_1}^{T_2} \frac{dT}{T^2} \tag{5.5}$$

Mathematical note: The negative sign arising from the integral of (dT/T) is accommodated by reversing the reciprocal terms in the brackets.

which leads to

$$\ln\left(\frac{P_2}{P_1}\right) = \frac{\Delta H^{vap}}{R}\left(\frac{1}{T_1} - \frac{1}{T_2}\right) \tag{5.6}$$

From a practical point of view, a useful form of eqn 5.6 can be obtained by using the fact that the *normal* boiling point is the temperature at which the vapour pressure is 1 atmosphere. Thus,

$$\ln\left(\frac{P}{1\ \text{atm}}\right) = \ln\left(\frac{P}{760\ \text{Torr}}\right) = \frac{\Delta H^{vap}}{R}\left(\frac{1}{T_{b,norm}} - \frac{1}{T}\right) \tag{5.7}$$

Example 5.2   The normal boiling point of benzene is 80.1 °C and the enthalpy of vaporization is 30.8 kJ mol^{-1}. Calculate the boiling point at a pressure of 100 torr.

Using eqn 5.7

$$\ln\left(\frac{100\ \text{mmHg}}{760\ \text{Torr}}\right) = \frac{30.8 \times 10^3\ \text{J mol}^{-1}}{8.314\ \text{J K}^{-1}\text{mol}^{-1}}\left(\frac{1}{273.15 + 80.1} - \frac{1}{T}\right)$$

Hence $(1/T) = 3.378 \times 10^{-3}$ K^{-1} or $T = 196.05$ K, 22.9 °C

An alternative approach is to take the indefinite integral of eqn 5.4, in which case we get

$$\ln P = (\text{constant}) - \frac{\Delta H^{\text{vap}}}{R}\left(\frac{1}{T}\right) \tag{5.8}$$

This is known as the *Clausius–Clapeyron equation*. Whereas eqn 5.3 suggests that the variation of melting point with temperature is a linear function, eqn 5.7 shows that this is not so where gas phases are involved. This can also be seen from the phase diagrams in Fig. 5.4.

Further progress can be made if, in eqn 5.7, we put P_1 equal to standard pressure, 1 bar. T_1 then becomes the *standard* boiling point, T_b. (We need to be careful to differentiate between the *standard* boiling point, with a vapour pressure of 1 bar and the *normal* boiling point where it is 1 atm.) Under these conditions we can use eqn 5.2,

$$\ln(1/\text{bar}) = (\text{constant}) - \frac{\Delta H^{\text{o,vap}}}{R}\left(\frac{1}{T_b}\right) \text{ or (constant) } = \frac{\Delta H^{\text{o,vap}}}{RT_b} = \frac{\Delta S^{\text{o,vap}}}{R}$$

which gives

$$\ln(P/\text{bar}) = \frac{\Delta S^{\text{o,vap}}}{R} - \frac{\Delta H^{\text{o,vap}}}{RT} \tag{5.9}$$

Thus, the entropy of vaporization can be measured from the intercept of a plot of $\ln(p/\text{bar})$ versus $(1/T)$. Some values are shown in Table 5.2. It is noticeable that many of the liquids listed have entropies of vaporization ≈ 85 J K^{-1} mol^{-1}. This empirical finding is known as *Trouton's rule*. Even liquids with considerably different chemical nature such as benzene, diethyl ether, and hydrogen sulfide do not deviate widely from this rule.

Trouton's rule can be rationalized since, in the absence of specific structural features, most liquids have similar structures. We know that most gases behave similarly. Therefore, it might be expected that the same change in structural order would take place on vaporization so that most liquids show similar entropy changes.

Deviations from Trouton's rule can usually be accounted for by specific features of the chemical bonding in the system. For example, water has a relatively ordered liquid structure owing to the hydrogen bonded network. Hence, there is a greater change of order on vaporization so that ΔS^{vap} is higher than expected. A compound exhibiting the opposite trend is acetic acid. Here, the vapour contains a substantial proportion of hydrogen bonded dimers so that there is a relatively ordered gas phase structure. Hence, ΔS^{vap} is lower than that predicted by Trouton's rule.

5.3 Phase behaviour in two component systems

As might be expected, the presence of the second component can complicate the phase behaviour. There can still be only one vapour phase, since gases mix in all proportions. As in a single component system, there can be any number of solid phases. The major difference is that now there can be more than one liquid phase. This is an everyday occurrence. For example, water and alcohol mix readily; water and petrol do not mix at all—we say they are *immiscible* in all proportions.

Note: Equation 5.8 can be expressed in a linear form by plotting ln *P* versus 1/*T*. This provides a convenient experimental method for measuring the enthalpy of vaporization.

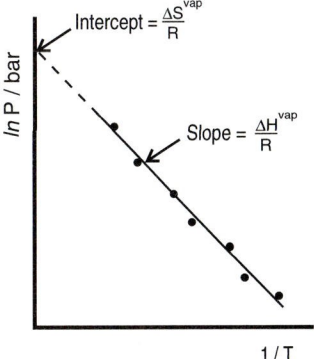

Mathematical note: ln (1) = 0.

Table 5.2 Standard entropy of vaporization (J K^{-1} mol^{-1}) for compounds at their boiling points

	T_b/K	$\Delta S^{\circ}{}^{\text{vap}}$
Benzene	353.2	87.2
CCl$_4$	349.3	85.9
n-Pentane	309.4	83.3
n-Decane	447.3	87.4
Cyclohexane	353.8	85.1
Hydrogen sulfide	212.8	87.9
Diethyl ether	307.8	84.5
Water	373.2	109.1
Acetic acid	391.0	59.8

Acetic acid is more correctly known as ethanoic acid.

Note: Acetic acid forms dimers in the gas phase

An alternative derivation of eqn 5.7 arises from treating this phase equilibrium according to the methods developed for reaction equilibrium in Chapter 4. Consider the vaporization as an equilibrium.

$$(liquid) \leftrightarrow (vapour)$$

We can write

$$\Delta G^{\circ vap} = \Delta H^{\circ vap} - T\Delta S^{\circ vap}$$
$$= \exp - (K_p/RT)$$

Formulating the expression for K_p

$$\ln K_p = \ln\left(\frac{P_{vap}}{1}\right) = \frac{-\Delta G^{\circ vap}}{RT}$$
$$= \frac{-(\Delta H^{\circ vap} - T\Delta S^{\circ vap})}{RT}$$

$$\ln(p/bar) = \frac{\Delta S^{\circ vap}}{R} - \frac{\Delta H^{\circ vap}}{RT}$$

which is the same as eqn 5.9.

Example 5.3   Many of the highest mountains in the world are in excess of 25 000 ft. At these altitudes the atmospheric pressure is in the region of 250 torr. Calculate the freezing point and boiling point of water at this pressure.

(The densities of ice and water at $0\ ^\circ C$ are 0.92 and 1.00 g cm^{-3} and the enthalpy of fusion of ice is 6.01 kJ mol^{-1}. The enthalpy of vaporization at the normal boiling point is 40.7 kJ mol^{-1}.)

The effect on the melting point is given by the Clapeyron equation in a manner analogous to Example 5.1.

$$\Delta V^{fus} = 18.01\text{ g mol}^{-1}\left(\frac{1}{1.00\text{ g cm}^{-3}} - \frac{1}{0.92\text{ g cm}^{-3}}\right)$$
$$= -1.57\text{ cm}^3\text{ mol}^{-1}$$
$$\frac{dP}{dT} = \frac{6.01 \times 10^3\text{ J mol}^{-1}}{(273.15\text{ K})(-1.57 \times 10^{-6}\text{ m}^3)} = -1.401 \times 10^7\text{ Pa K}^{-1}$$
$$\Delta P = (760 - 250) = 510\text{ torr} = 510 \times 133.32\text{ Pa torr}^{-1}$$
$$= 6.80 \times 10^4\text{ Pa}$$

Hence $\Delta T = 6.80 \times 10^4\text{ Pa}/(-1.401 \times 10^7\text{ Pa K}^{-1}) = 4.85 \times 10^{-3}\text{ K}$

Thus, the freezing point is $-0.005\ ^\circ C$.

The effect on the boiling point is given by the Clausius–Clapeyron equation. Using eqn 5.7 with $P_1 = 1\text{ atm} = 760\text{ torr}$ and $T_1 = 100\ ^\circ C$, leads to

$$\ln\left(\frac{250}{760}\right) = \frac{+40.7 \times 10^3\text{ J mol}^{-1}}{8.314\text{ J K}^{-1}\text{mol}^{-1}}\left(\frac{1}{373.15\text{ K}} - \frac{1}{T_b/K}\right)$$
$$T_b = 343.99\text{ K} = 70.8^\circ C$$

The results demonstrate that the melting point is influenced to a much lesser degree than the boiling point. The difference in the melting point is barely noticeable except with the most accurate thermometer. The difference in boiling point would be obvious to a cold mountaineer! For this reason, tables of boiling points of compounds usually report the pressure at which the measurement is made. This is rarely done for melting points.

Liquid–liquid phase behaviour

Consider two liquids, A and B, at constant pressure and temperature which have molar Gibbs free energies of G_A and G_B, respectively. If we mix n_A moles of A with n_B moles of B then

$$\text{Gibbs free energy } before \text{ mixing} = (n_A\ G_A + n_B\ G_B)$$
$$\text{Gibbs free energy } after \text{ mixing} = (n_A + n_B)\ G_{(A+B)}$$

where $G_{(A+B)}$ is the molar Gibbs free energy of the mixture. We define the Gibbs free energy of mixing, ΔG^{mix} as the difference in G after mixing from that before. Hence,

$$\Delta G^{mix} = [(n_A + n_B)G_{(A+B)}] - [n_A G_A + n_B G_B]$$

Dividing by the total number of moles, $(n_A + n_B)$, then

$$\Delta G^{mix} = [G_{(A+B)}] - [x_A G_A + x_B G_B] \qquad (5.10)$$

As we have argued in other areas, the condition that must be satisfied if mixing is to occur is that ΔG^{mix} must be negative. The system will come to equilibrium when G is at its most negative value. Three possibilities exist and these are illustrated in Fig. 5.6 which shows schematically how ΔG^{mix} can vary with the mole fraction of the mixture.

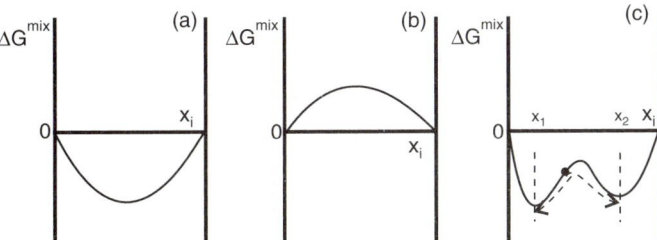

Fig. 5.6 Gibbs free energy of mixing for two liquids. See text for explanation.

In Fig. 5.6(a), the Gibbs free energy of mixing is negative for all possible compositions so that the two liquids mix in all proportions at the temperature involved. They are completely *miscible*. Conversely, in (b), ΔG^{mix} is positive throughout the composition range so that no mixing can occur leading to immiscibility. The situation in (c) is more interesting. ΔG^{mix} is negative for all compositions and so a miscible mixture would be expected. However, if the mixture has a composition between x_1 and x_2 then a less negative Gibbs free energy will result if the mixture separates into two phases with compositions x_1 and x_2. This type of system is described as *partially miscible*. The liquids are miscible for mole fractions less than x_1 or greater than x_2 and immiscible for mole fractions between these values. Here, a negative Gibbs free is a necessary, but not sufficient, condition for miscibility.

To understand how partial miscibility can arise, we need to consider the contributions from the enthalpy and entropy of mixing. These are defined by equations analogous to eqn 5.10.

$$\Delta H^{mix} = (H_{(A+B)}) - (x_A H_A + x_B H_B); \; \Delta S^{mix} = (S_{(A+B)}) - (x_A S_A + x_B S_B)$$

In the vast majority of cases (we will discuss the exceptions shortly), ΔS^{mix} will be positive since the mixture must be more disordered (higher entropy) than the pure components. The $-T\Delta S^{mix}$ contribution to the free energy will therefore be favourable and hence it is the enthalpic contribution that largely determines the miscibility. The possibilities are illustrated in Fig. 5.7.

In Fig. 5.7(a), the mixing is exothermic so ΔH^{mix} is negative. Both contributions are therefore favourable and the components are miscible in all proportions. In (b), mixing is endothermic and $\Delta H^{mix} > -T\Delta S^{mix}$ so that ΔG^{mix} is positive and no mixing occurs at any composition. Mixing is also endothermic in (c) but here $\Delta H^{mix} < -T\Delta S^{mix}$ so that the resultant free energy is negative and the system is miscible. Case (d) is a special case of case (c) and occurs when the concentration dependence of ΔH^{mix} has a particular form resulting in the free energy function shown. This can be summarized as:

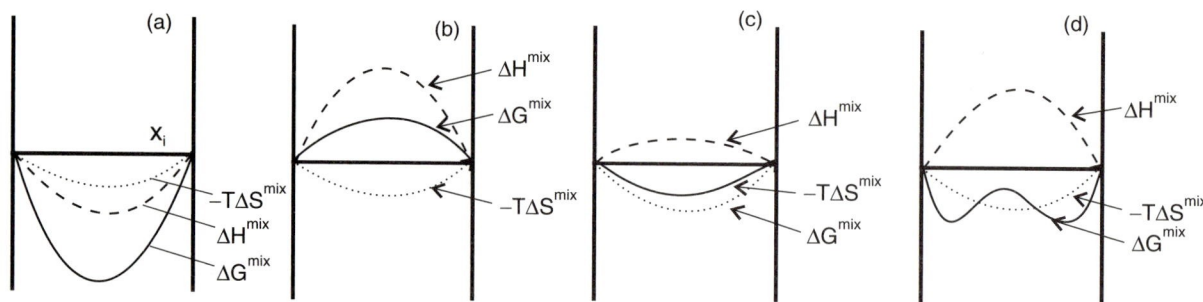

Fig. 5.7 The contributions to the Gibbs free energy of mixing.

1. S^{mix} + ve; ΔH^{mix} negative; $\qquad$ ΔG^{mix} highly negative $\quad \therefore$ miscible
2. S^{mix} + ve; ΔH^{mix} + ve $> -T\Delta S^{mix}$; $\quad \Delta G^{mix}$ positive $\qquad \therefore$ immiscible
3. S^{mix} + ve; ΔH^{mix} + ve $< -T\Delta S^{mix}$; $\quad \Delta G^{mix}$ small, negative $\quad \therefore$ miscible
4. S^{mix} + ve; ΔH^{mix} + ve $\approx -T\Delta S^{mix}$; $\quad \Delta G^{mix}$ highly negative

$$\therefore \text{ partially miscible}$$

However, that is not the whole story and, as so often in thermodynamics, we can introduce another layer of complexity. We have so far been considering the systems at constant temperature. Since T appears in the entropic term, we must consider the consequences of changing the temperature.

In cases 1 and 3, a change of T will merely change the magnitude of the negative free energy but can never change its sign. Therefore, such a system will be miscible at all temperatures.

In case 2, the entropy term will become dominant at high temperatures. Therefore, we might expect a system to be immiscible at low temperatures but to become miscible on heating. We call the maximum temperature at which this happens the *upper critical solution temperature*, UCST. This is shown schematically in Fig. 5.8. A similar argument applies to case 4 where we would expect increasing compatibility of the mixture as the temperature was increased.

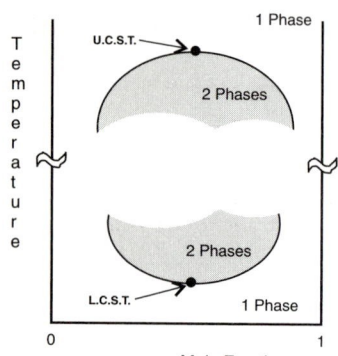

Fig. 5.8 Schematic representation of Upper (UCST) and lower (LCST) critical solution temperatures. The system changes from two phase behaviour (shaded) to one phase (non-shaded).

Note: A further level of complication can arise since the enthalpy and entropy of mixing are themselves temperature dependent and so can switch sign. A complete description is outside the scope of this primer. Such cases are rare but can lead to a mixture showing both UCST and LCST behaviour. Such a system is nicotine–water shown in Fig. 5.9.

It was stated above that ΔS^{mix} would be positive for the majority of mixtures. The exception occurs where the two components have some kind of strong interaction, such as hydrogen bonding, which results in the mixture having a *higher* degree of order than the pure components. This means that the entropic contribution is unfavourable to mixing but this is almost always overcome by a high ΔH^{mix} associated with the strong interactions. In this case, the entropy term favours *demixing* or *phase separation* as the temperature is increased. The lowest temperature at which this occurs is the *lower critical solution temperature*, LCST (see Fig 5.8).

Again, much of the information describing phase behaviour can be represented on a phase diagram. With two components, however, we have three variables—pressure, temperature, and composition. To avoid the need for drawing three-dimensional diagrams, we normally cheat somewhat and work at constant pressure so that we use temperature–composition diagrams to describe the phase behaviour. Some examples are shown in Fig. 5.9.

Vapour–liquid phase behaviour

We now turn to a discussion of what factors influence the vapour in equilibrium with a liquid mixture. It turns out that this yields a great deal of

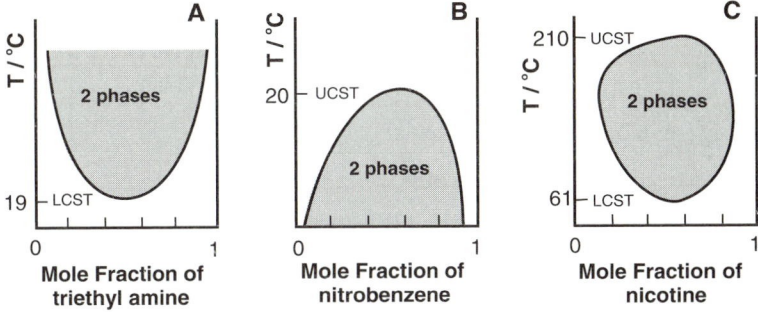

Fig. 5.9 Phase diagrams for two component liquid mixtures. The two-phase regions are shaded.

A: Water–triethyl amine. B: Hexane–nitrobenzene. C: Water–nicotine.

Exercise: Consider the chemical nature and structures of the compounds involved in the phase diagrams in Fig. 5.9. Can you explain the behaviour observed in each case in the light of the discussion in the previous section?

useful information on the structure and properties of the solution. Perhaps the main application of vapour–liquid equilibrium, VLE, has been in describing distillation processes. Although now considered largely the province of chemical engineers, distillation remains the major, large-scale separation method used in the chemical industry. A detailed discussion of distillation is not appropriate for this primer and we will concentrate on what VLE can tell us about the solution properties.

We start with an empirical observation. Working towards the end of the 19th century, Francois Raoult reported that, for a large number of solutions, the vapour pressure exerted by a component (its *partial vapour* pressure) was proportional to its mole fraction concentration. Thus, for a mixture of components A and B,

$$p_A = x_A p_A^\circ; \quad p_B = x_B p_B^\circ \qquad (5.11)$$

and
$$p_{\text{TOTAL}} = p_A + p_B = x_A p_A^\circ + x_B p_B^\circ \qquad (5.12)$$

In eqns 5.11 and 5.12, p° represents the vapour pressure of the pure component, sometimes called the *saturated vapour pressure* at the temperature of the solution. This can be seen in Fig. 5.10. Equation 5.11 has come to be known as *Raoult's law*. In fact, we now know that it is, at best, an approximation, but it works quite well for mixtures of similar compounds such as hexane–heptane or benzene–toluene. We will shortly see the reason for this.

How can we treat this situation in terms of the thermodynamic principles already developed? Consider two containers as shown in Fig. 5.11 which are held at the same temperature and allowed to come to equilibrium. Container 1 contains pure liquid A so that the vapour pressure inside must be p_A°. The second container has a solution of A and B with mole fractions x_A and x_B. For simplicity, here we will only consider *non-electrolyte* systems so that we do not have to worry about the presence of ions or the extent of ionization of the compounds.

From the requirement for phase equilibrium developed in Section 5.2, we know that the chemical potential of A has to be the same in each phase. We can therefore consider μ_A under each of the conditions:

From 1. $\mu_A(\text{gas}, p_A^\circ) = \mu_A(\text{pure liquid A}) \qquad (5.13)$

From 2. $\mu_A(\text{gas}, p_A) = \mu_A(\text{mixture at } x_A) \qquad (5.14)$

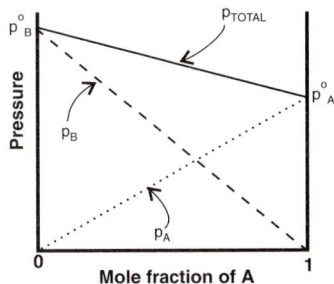

Fig. 5.10 Variation of pressure over a solution obeying Raoult's law.

Note: Don't forget that, for a binary mixture, $x_A + x_B = 1$.

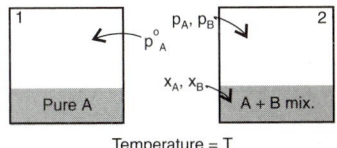

Fig. 5.11 Relationship between solution properties and those of the pure components.

From Section 4.6, we know that

$$\mu_A(\text{gas}, p_A^\circ) = \mu_A(\text{gas}, 1 \text{ bar}) + RT \ln(p_A^\circ/\text{bar}) \tag{5.15}$$
$$\mu_A(\text{gas}, p_A) = \mu_A(\text{gas}, 1 \text{ bar}) + RT \ln(p_A/\text{bar}) \tag{5.16}$$

Subtracting eqn 5.16 from 5.15 yields

$$\mu_A(\text{gas}, p_A) - \mu_A(\text{gas}, p_A^\circ) = RT \ln(p_A/p_A^\circ) \tag{5.17}$$

Substituting eqn 5.13 and 5.14 into 5.17 gives

$$\mu_A(\text{mixture at } x_A) - \mu_A(\text{pure liquid A}) = RT \ln(p_A/p_A^\circ)$$

If we define the reference state as the pure liquid (as previously) with chemical potential μ_A°, and the chemical potential of the component in solution is μ_A, then

$$\mu_A = \mu_A^\circ + RT \ln(p_A/p_A^\circ) \tag{5.18}$$

Note that the only assumptions made here are in eqns 5.15 and 5.16 in assuming ideal gas behaviour for the vapour. It makes no assumption of any particular properties for the mixture.

Ideal solutions

Note: Many people define an ideal solution as one which obeys Raoult's Law. Strictly, this is not a definition but an experimental consequence of the definition which is given by eqn 5.19.

Equation 5.18 is completely general. If we apply it to solutions where Raoult's Law is obeyed, then it can be modified to

$$\mu_A = \mu_A^\circ + RT \ln x_A \tag{5.19}$$

As in the case of ideal gases, it is useful to define an *ideal solution* to use as a first approximation to solution behaviour. In this case, we define an ideal solution as one that obeys eqn 5.19. In practice, this means that it is one that obeys Raoult's law.

In order to account for the properties of ideal solutions, we need to consider why they display a linear dependence of vapour pressure with concentration. This implies that the partial pressure exerted by a component depends only on its concentration and does not depend on the other component, which has only a dilution effect. This explains the observation that solutions of chemically similar compounds display ideal behaviour. The intermolecular interactions will be the same in solution as in the pure compounds. Schematically, this is represented in Fig. 5.12. In an ideal solution, the environment experienced by the central A molecule (circled) will be the same in pure A, pure B, or in a solution of A and B. If the interactions are the same, it implies that there will be a zero enthalpy change on mixing A and B.

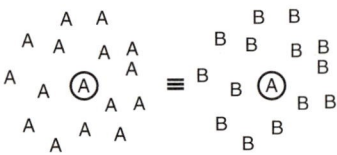

Fig. 5.12 Schematic representation of an ideal solution. The circled molecules have the same intermolecular interaction in both cases.

There are also entropic considerations. Formation of an ideal solution implies that the mixing is random. We can quantify this by taking the integral form of eqn 4.26 to give the total Gibbs free energy of a solution in terms of the chemical potentials.

$$G_{p,T} = \sum_i \mu_i n_i$$

For 1 mole of solution, n_i is replaced by the mole fraction of the component, x_i. Hence, the Gibbs free energy of mixing is given by

$$\Delta G^{\text{mix}} = G(\text{solution}) - G(\text{pure components})$$
$$\Delta G^{\text{mix}} = \sum_i \mu_i x_i - \sum_i \mu_i^\circ x_i = \sum_i x_i(\mu_i^\circ + RT \ln x_i) - \sum_i \mu_i^\circ x_i$$

The terms in μ° cancel to leave

$$\Delta G^{\mathrm{mix}} = RT \sum_i x_i \ln x_i \qquad (5.20)$$

The expression for the entropy of mixing can be obtained by using the usual relation

$$\Delta G^{\mathrm{mix}} = \Delta H^{\mathrm{mix}} - T\Delta S^{\mathrm{mix}} = 0 - T\Delta S^{\mathrm{mix}}$$

from which it is apparent that

$$\Delta S^{\mathrm{mix}} = -R \sum_i x_i \ln x_i \qquad (5.21)$$

so that the entropy of mixing can be found merely from the mole fraction concentrations of the components.

Thus, we can identify two requirements at the molecular level for compounds to form an ideal solution. They should consist of molecules:

- which are similar in chemical nature so that they have similar intermolecular forces; and
- which are similar in size and shape.

The second of these is important in satisfying the requirement for random mixing. The requirement for molecules to have a similar size and shape suggests there will be no overall volume change on mixing. This can be shown by considering eqn 4.10, in which we showed that the volume of a system depended on the pressure dependence of the Gibbs free energy. Applying the equation to this case

$$\Delta V^{\mathrm{mix}} = \left(\frac{\partial \Delta G^{\mathrm{mix}}}{\partial P} \right)_T = \left(\frac{\partial RT \sum_i x_i \ln x_i}{\partial P} \right)_T = 0 \qquad (5.22)$$

Note: In practice, the first requirement is usually more important, except in the case of solutions of polymer molecules, where the size difference between polymer and solvent can be very large.

Note: Many people do not realize that, in general, liquid volumes are not additive. As an example a mixture of 50 cm³ of water and 50 cm³ ethanol will give 97 cm³ of solution. The contraction arises since the presence of ethanol disrupts the hydrogen bonded structure of the water (and vice versa) so that the molecules occupy different volumes in the solution than in the pure liquids.

Example 5.4 Calculate the changes in enthalpy, entropy, volume, and Gibbs free energy on mixing 1.5 mol of hexane with 2.5 mol heptane at 25 °C.

This mixture can be assumed to be ideal. Therefore $\Delta H^{\mathrm{mix}} = \Delta V^{\mathrm{mix}} = 0$. The other functions are given by eqns 5.20 and 5.21.

The mole fraction of hexane is $1.5/(1.5 + 2.5) = 0.375$. $x_{\mathrm{heptane}} = 0.625$

$$\Delta G^{\mathrm{mix}} = (8.314)(298.15)\{0.375 \ln (0.375) + 0.625 \ln (0.625)\}$$
$$= -1639.9 \text{ J mol}^{-1}$$
$$\Delta S^{\mathrm{mix}} = -(8.314)\{0.375 \ln (0.375) = 0.625 \ln (0.625)\}$$
$$= +5.5 \text{ J K}^{-1} \text{ mol}^{-1}$$

Note that the formation of the solution is 'driven' by the greater disorder in the mixture over that of the pure liquids.

We have commented several times that the ideal gas state is useful since most gases follow it under conditions in which we normally encounter them. The ideal solution model is less useful in that few solutions display ideal behaviour. In fact, Raoult's law is, at best, an approximation. What use then is the ideal solution model? It has allowed us to define a solution in terms of the molecular properties of its constituents. In fact, it is in *deviations* from ideal solution behaviour where the usefulness of the ideal model lies in that it enables us to say something about the chemical interactions involved.

Non-ideal solutions

Once again, in order to use the same equations we have already developed, we use the concept of thermodynamic activity (see Section 4.5). Equation 5.18 is therefore written as

$$\mu_A = \mu_A^\circ + RT \ln a_A \qquad (5.23)$$

where the activity, a_A, is given by (p_A/p_A°). In an ideal solution, the activity corresponds to the mole fraction of the component in the solution. For real solutions, this is modified by the addition of an *activity coefficient*, γ, where

$$a_A = (p_A/p_A^\circ) = \gamma_A x_A \qquad (5.24)$$

It follows from eqn 5.24 that

$$\gamma_i = \frac{a_i}{x_i} = \frac{\text{'effective concentration'}}{\text{'actual concentration'}}; = \frac{p_i}{x_i p_i^\circ} = \frac{p_i(\text{real})}{p_i(\text{ideal})}$$

or

$$p_i = \gamma_i x_i p_i^\circ = \gamma_i p_i^{\text{ideal}} \qquad (5.25)$$

so that γ tells us how the actual properties of the solution differ from those of the ideal model we proposed in the previous section. The larger the value of γ, the greater the deviation from ideal behaviour. Let us consider two examples.

The ideal model arises when the intermolecular interactions between the components are identical. In the vast majority of cases, the interactions in solution are weaker than those in the pure liquids. In this case, molecules are 'held less tightly' into the solution and can more readily escape into the vapour. Thus, the actual vapour pressure is greater than would be the case if the solution conformed to ideal behaviour. The vapour pressure curves have the form shown in Fig. 5.13. Such systems are said to display *positive deviations from Raoult's law*. It is also clear that $p_i > p_i^{\text{ideal}}$ so that $\gamma_i > 1$.

In the opposite case, some systems show quite strong intermolecular interactions in solution. A common example is that the two components can form hydrogen bonds, although this is not possible in the pure liquids. Here, the molecules are 'held more tightly' in the solution so that less escape to the vapour so that $p_i < p_i^{\text{ideal}}$ and $\gamma_i < 1$. The vapour pressure curve takes the form shown in Fig. 5.14. The best known example of this behaviour is in mixtures of acetone and chloroform. Both liquids have relatively simple structures but in the mixture hydrogen bonding is possible between the dipoles of the C–H in $CHCl_3$ and the C=O in the carbonyl group in acetone, as shown in Fig. 5.15. Such systems are said to display *negative deviations from Raoult's law*.

More sophisticated treatments than can be outlined here relate γ to Gibbs free energy changes that occur in the system and can be used to characterize extensively the interactions that take place in solution. While the discussion here has centred on mixtures of non-electrolytes, models which account for the very non-ideal behaviour in solutions where ionization occurs are also available.

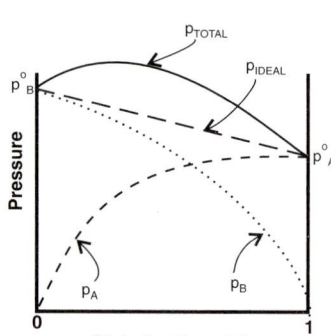

Fig. 5.13 Vapour pressure curves for systems exhibiting positive deviations from Raoult's law.

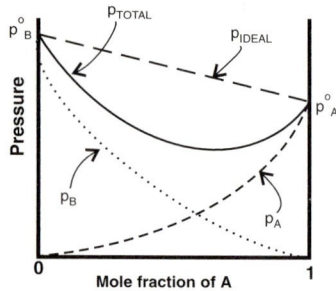

Fig. 5.14 Vapour pressure curves for systems exhibiting negative deviations from Raoult's law.

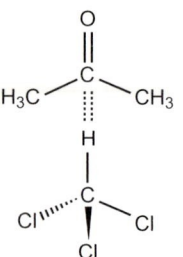

Fig. 5.15 Interactions in solutions of acetone and chloroform.

Note: A fuller discussion of the properties of non-ideal solutions can be found in the texts listed in Further Reading.

Example 5.5 A solution of 16.6 g ethanol (C_2H_5OH) and 65.5 g of methyl cyclohexane ($C_6H_{11}CH_3$) were placed in a sealed container at 55 °C. The vapour pressure over the solution, which had a mole fraction of ethanol of 0.523, was 50.13 kPa. Saturated vapour pressures and relative molar masses for ethanol are 37.33 kPa and 46.07 g mol^{-1} and for methyl cyclohexane are 22.40 kPa and 98.19 g mol^{-1}. Calculate the activity coefficients for both components.

We first need to calculate the mole fractions of the liquids in solution

$$x_{C_2H_5OH} = \frac{\text{No. of moles ethanol}}{\text{Total number of moles}} = \frac{\left(\frac{16.6}{46.07}\right)}{\left(\frac{16.6}{46.07}\right) + \left(\frac{65.5}{98.19}\right)} = 0.350$$

so that $x_{C_7H_{14}} = 0.650$.

To calculate the activity coefficients, we use eqn 5.25, $p_i = \gamma_i x_i p_i^\circ$, but first need to calculate the partial pressure of each component in the vapour phase. This is done using Dalton's law, which states that the partial pressure is the mole fraction in the vapour phase multiplied by the total pressure.

$$p_{C_2H_5OH} = (\text{vapour mol. frac.}) \times (p_{\text{total}}) = 0.523 \times 51.13 = 26.218 \text{ kPa}$$
$$p_{C_7H_{14}} = (1 - 0.523) \times 51.13 = 24.389 \text{ kPa}$$

Hence,

$$p_{C_2H_5OH} = \gamma_{C_2H_5OH} \, x_{C_2H_5OH} \, p_{C_2H_5OH}^\circ; \quad p_{C_7H_{14}} = \gamma_{C_7H_{14}} \, x_{C_7H_{14}} \, p_{C_7H_{14}}^\circ$$
$$26.218 \text{ kPa} = \gamma_{C_2H_5OH}(0.350)(37.33 \text{ kPa}); \quad \gamma_{C_2H_5OH} = 2.01$$
$$24.389 \text{ kPa} = \gamma_{C_7H_{14}}(0.650)(22.40 \text{ kPa}); \quad \gamma_{C_7H_{14}} = 1.67$$

Note that, as expected for components with dissimilar chemical nature, both activity coefficients are > 1 indicating positive deviations from Raoult's law. ($p^{\text{ideal}} = 32.10$ kPa). The non-polar methyl cyclohexane disrupts the hydrogen bonding in ethanol.

Solid–liquid phase equilibrium

As the final stage of our discussion of phase equilibrium in two component systems, we need to consider briefly the melting and freezing transition. At the melting point (or freezing point) we have a solid in equilibrium with the liquid mixture.

Consider a completely miscible mixture of two non-electrolytes at constant pressure. Conventionally, we call the component present in the larger amount the *solvent* and the other component the *solute*. For convenience, we will label the solute A and solvent B, although in thermodynamic terms no differentiation necessary is necessary. On cooling the mixture, solidification will take place. However, from everyday experience there are two distinct possibilities:

- *solid A ↔ liquid (A + B)*. Solid A comes out of solution leaving a saturated solution of A in B. This represents the solubility limit of the solute in the solvent.

• *solid B ↔ liquid (A + B)*. The solvent starts to freeze. This represents the freezing point of the solution.

We can use the same thermodynamic model to treat both of these situations. Consider component A and the equilibrium:

$$\text{solid A} \leftrightarrow \text{A in solution}$$

Note: The superscript ° reminds us that these are pure component properties. μ_A (solution) is given in eqn 5.26 by eqn 4.26.

At equilibrium, the chemical potentials must be equal so that

$$\mu_A^\circ(\text{solid}) = \mu_A(\text{solution}) = \mu_A^\circ(\text{liquid}) + RT \ln a_A \qquad (5.26)$$

or
$$\mu_A^\circ(\text{solid}) - \mu_A^\circ(\text{liquid}) = RT \ln a_A \qquad (5.27)$$

The left hand side of eqn 5.27 is the difference in molar Gibbs free energy between the pure solid and pure liquid at the melting point, T_m, i.e. the free energy of fusion, ΔG^{fus}, which as usual can be split into enthalpic and entropic contributions.

$$\Delta G_A^{\text{fus}} = \Delta H_A^{\text{fus}} - T\Delta S_A^{\text{fus}} = RT \ln a_A \qquad (5.28)$$

We recall from Section 5.2 that, for a phase change at the transition temperature, $\Delta S_A^{\text{fus}} = (\Delta H_A^{\text{fus}}/T_{m,A})$ where $T_{m,A}$ is the melting point of pure A. Substituting this into eqn 5.28 gives

$$\Delta H_A^{\text{fus}} - T(\Delta H_A^{\text{fus}}/T_{m,A}) = RT \ln a_A$$

Note: Strictly, eqn 5.29 is valid only for temperatures near the pure component melting point. Also, the temperature dependence of ΔH_A^{fus} should be taken into account. However, these effects are usually small and can safely be ignored.

which can be rearranged to give

$$\ln a_A = \frac{\Delta H^{\text{fus}}}{R}\left(\frac{1}{T_{m,A}} - \frac{1}{T}\right) \qquad (5.29)$$

For an ideal solution, $a_A = x_A$ so that substituting this into eqn 5.29 gives the *ideal solubility equation*. This equation tells how the melting point, T, of a solution varies with composition, a_A. The same treatment could be applied to B leading to the same equation with all the pure component terms now applying to B.

One noticeable feature of eqn 5.29 is that it contains only terms involving component A. The consequence of this is that, at any temperature T, the solubility of A should be the same (if measured on a mole fraction basis) in any solvent. This is a good test of whether a solution behaves ideally, since, if it forms a non-ideal solution, the activity coefficient will vary with different solvents.

Note: The structures of the compounds are:

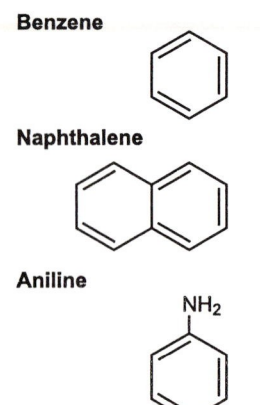

Benzene

Naphthalene

Aniline

NH₂

Example 5.6   The solubility of naphthalene at 20 °C is 24 mol percent in benzene and 13 mole percent in aniline (amino benzene). Pure naphthalene melts at 80.2 °C and has an enthalpy of fusion of 19.05 kJ mol⁻¹. Comment.

We can make a prediction of the solubility from the ideal solubility equation, Equation 5.29 with 'A' representing naphthalene and the activity given by the mole fraction.

$$\ln x_A = \frac{\Delta H^{\text{fus}}}{R}\left(\frac{1}{T_{m,A}} - \frac{1}{T}\right) = \frac{19050 \text{ J mol}^{-1}}{8.314 \text{ mol}^{-1}\text{K}^{-1}}\left(\frac{1}{353.4} - \frac{1}{293.2}\right)\text{K}^{-1}$$

which leads to a mole fraction $x_{\text{napththalene}} = 0.264$ or 26.4 mol%.

The solubility in benzene is close to the ideal value, as would be expected from the similar chemical structures. Aniline is more polar and has other interactions and so behaves as a non-ideal solution. The solubility is significantly different from the ideal value.

Note that we can calculate the activity coefficients for napththalene in both solvents to quantify the deviation from ideal behaviour. For benzene, $\gamma = 1.10$ (close to the ideal value of 1.0) and for aniline $\gamma = 2.03$.

5.4 Concluding remarks

This chapter has introduced some of the methods used to describe the phase behaviour of compounds. Only relatively straightforward systems with one or two components have been covered, but the treatment of more complex systems with any number of components and any number of phases builds on the ideas used here. The work described here should suffice to convince you that consideration of straightforward principles allows us to understand a number of aspects of the phase behaviour of compounds. The guiding tenet in determining phase behaviour is once again the tendency of a system to attain a state of minimum free energy (Gibbs free energy at constant pressure). In this, the principles involved in the study of phase equilibrium parallel those of chemical reaction equilibrium.

The overall aim of this primer was to introduce the concepts and ideas of energy changes during chemical processes and reactions and how these ideas can be used to correlate and explain chemical reactivity and phase behaviour. I hope that after reading and working through the book and the examples, you will be in a position to perform calculations and predictions on the course of reactions and the phase behaviour of compounds. A small range of problems is given in Appendix 2 for you to try and you would be well advised to try a large number of others. Good luck.

Appendix 1: thermochemical data

This brief list has been constructed to allow students to tackle the problems given in Appendix 2, and to allow instructors to set further problems. Sources: *Physical chemistry*, 5th edn, P. W. Atkins, OUP 1995 and *Handbook of chemistry and physics*, 61st Edn, R. C. Weast (Ed.), CRC Press 1981.

All data refer to 298.15 K and 1 bar pressure. Units of $\Delta H°$ and $\Delta G°$ are kJ mol^{-1}; Units of $S°$ and c_p are J K^{-1} mol^{-1}

Compound		$\Delta H°$	$\Delta G°$	$S°$	c_p
Carbon	$C_{(graphite)}$	0	0	5.7	8.5
	$C_{(diamond)}$	+1.9	+2.9	2.4	6.1
Carbon monoxide	$CO_{(g)}$	−110.5	−137.2	197.7	29.1
Carbon dioxide	$CO_{2(g)}$	−393.5	−394.4	213.7	37.1
Methane	$CH_{4(g)}$	−74.8	−50.7	186.3	35.3
Ethane	$C_2H_{6(g)}$	−84.7	−32.8	229.6	52.6
Ethene	$C_2H_{4(g)}$	+52.3	+68.2	219.6	43.6
Ethyne	$C_2H_{2(g)}$	+226.7	+209.2	200.9	43.9
Propane	$C_3H_{8(g)}$	−103.9	−23.5	269.9	73.5
Butane	$C_4H_{10(g)}$	−126.2	−17.0	310.2	97.5
Pentane	$C_5H_{12(g)}$	−146.4	−8.2	348.4	120.2
Hexane	$C_6H_{14(l)}$	−198.7	0.2	204.3	142.1
Heptane	$C_7H_{16(l)}$	−224.4	+1.0	328.6	224.3
Cyclohexane	$C_6H_{12(l)}$	−156.0	+26.8		156.5
Benzene	$C_6H_{6(g)}$	+82.9	+129.7	269.3	81.7
Benzene	$C_6H_{6(l)}$	+49.0	+124.3	173.3	136.1
Toluene	$C_6H_5CH_{3(l)}$	+50.0	+122.0	320.7	103.6
Methanol	$CH_3OH_{(l)}$	−238.7	−166.3	126.8	81.6
Methanol	$CH_3OH_{(g)}$	−200.7	−161.9	239.8	43.9
Ethanol	$C_2H_5OH_{(l)}$	−277.7	−174.8	160.7	111.5
Ethanol	$C_2H_5OH_{(g)}$	−235.1	−168.5	282.7	65.4
Phenol	$C_6H_5OH_{(s)}$	−165.0	−50.9	146.0	
Acetic acid	$CH_3COOH_{(l)}$	−484.5	−389.9	159.8	124.3
Benzoic acid	$C_6H_5COOH_{(s)}$	−385.1	−245.3	167.6	146.8
Sucrose	$C_{12}H_{22}O_{11(s)}$	−2222	−1543	360.2	
β-D-glucose	$C_6H_{12}O_{6(s)}$	1268	−910	212	
Glycine	$CH_2(NH_2)CO_2H_{(s)}$	−532.9	−373.4	103.5	99.2
Urea	$CO(NH_2)_{2(s)}$	−333.5	−197.3	104.6	93.1
Bromine	$Br_{2(l)}$	0	0	152.2	75.7
Bromine	$Br_{2(g)}$	+30.9	+3.1	245.5	36.0
Hydrogen bromide	$HBr_{(g)}$	−36.4	−53.5	198.7	29.1
Calcium	$Ca_{(s)}$	0	0	41.4	25.3

Compound		$\Delta H°$	$\Delta G°$	$S°$	c_p
Calcium oxide	$CaO_{(s)}$	−635.1	−604.0	39.8	42.8
Calcium carbonate (*calcite*)	$CaCO_{3(s)}$	−1206.9	−1128.8	92.9	81.9
Tetrachloromethane	$CCl_{4(l)}$	−135.4	−65.2	216.4	131.8
Hydrogen cyanide	HCN	+135.1	+124.7	201.8	35.9
Chlorine	$Cl_{2(g)}$	0	0	223.1	33.9
Hydrogen chloride	$HCl_{(g)}$	−92.3	−95.3	186.9	29.1
Fluorine	$F_{2(g)}$	0	0	202.8	31.3
Hydrogen fluoride	$HF_{(g)}$	−271.1	−273.2	173.8	29.1
Hydrogen	$H_{2(g)}$	0	0	130.7	28.8
Water	$H_2O_{(l)}$	−285.8	−237.1	69.9	75.3
Water vapour	$H_2O_{(g)}$	−241.8	−228.6	188.3	33.6
Iron oxide	$Fe_2O_{3(s)}$	−824.2	−742.2	87.4	103.9
Iron oxide	$FeO_{(s)}$	−266.3	244.1	53.9	
Magnesium	Mg	0	0	32.7	24.9
Magnesium oxide	$MgO_{(s)}$	−601.7	−569.4	26.9	37.2
Magnesium carbonate	$MgCO_{3(s)}$	−1095.8	−1012.1	65.7	75.5
Nitrogen	$N_{2(g)}$	0	0	191.6	29.1
Ammonia	$NH_{3(g)}$	−46.1	−16.5	192.5	35.1
Nitrogen dioxide	$NO_{2(g)}$	+33.2	+51.3	240.1	37.2
Nitric oxide	$NO_{(g)}$	+90.3	+86.6	210.8	29.8
Dinitrogen tetroxide	$N_2O_{4(g)}$	+9.2	+97.9	304.3	77.3
Iodine	$I_{2(s)}$	0	0	116.1	54.4
Iodine	$I_{2(g)}$	+62.4	+19.3	260.7	36.9
Hydrogen iodide	$HI_{(g)}$	+26.5	+1.7	206.6	29.2
Oxygen	$O_{2(g)}$	0	0	205.1	29.4
Ozone	$O_{3(g)}$	+142.7	+163.2	238.9	39.2
Sulphur (rhombic)	$S_{(s)}$	0	0	31.8	22.6
Sulphur (monoclinic)	$S_{(s)}$	+0.3	+0.1	32.6	23.6
Sulphur dioxide	$SO_{2(g)}$	−269.8	−300.2	248.2	39.9
Sulphur trioxide	$SO_{3(g)}$	−395.7	−371.1	256.8	50.7

Appendix 2: additional problems

You should make use of data listed in Appendix 1 where needed.

1. An ideal gas expands reversibly and isothermally from 20 bar to 1 bar. Calculate the values (per mole) of q, w, ΔU, ΔH, ΔS, and ΔG.

2. The decomposition at constant volume of 1 mole of gaseous krypton difluoride, KrF_2, at 25 °C evolves 59.4 kJ mol^{-1} of heat. The enthalpy of sublimation of solid KrF_2 is 41 kJ mol^{-1}. Calculate $\Delta H^{\circ}_{f,298}$ for solid KrF_2.

3. When solid biphenyl, $(C_6H_5)_2$ is burned at 25 °C in a bomb calorimeter in excess oxygen, it was completely converted to gaseous CO_2 and liquid water with the evolution of 6243 kJ mol^{-1} of heat. Use standard enthalpies of formation for the products to calculate $\Delta H^{\circ}_{f,298}$ for biphenyl.

4. The standard enthalpy change for the reaction $N_{2(g)} + 3H_{2(g)} \rightarrow 2\,NH_{3(g)}$ is -92.4 kJ $mol(N_2)^{-1}$ at 300 K. Estimate the enthalpy change for the reaction at 800 K. What assumptions do you make?

5. Calculate the entropy change for 3 mol of hydrogen gas at 1 bar on heating from 0 to 500 °C given that the heat capacity varies with temperature according to : $c_p(T) = 27.3 + 3.8 \times 10^{-3}$ T J K^{-1} mol^{-1}.

6. An equimolar mixture of carbon monoxide and hydrogen is allowed to come to equilibrium over a catalyst which promotes their conversion to methanol, gaseous CH_3OH. At an equilibrium pressure, P, the fraction of CO converted to methanol is X. Formulate an expression for K_p as a function of P and X and show that, for very low conversions, the conversion is proportional to the square of the pressure. Use the standard Gibbs free energies to calculate the value of K_p for this reaction.

7. At high temperatures, ethane decomposes into ethene and hydrogen according to

$$C_2H_{6(g)} \rightarrow C_2H_{4(g)} + H_{2(g)}$$

Calculate the changes in enthalpy, entropy and Gibbs free energy for the reaction at 1200 K and 1 bar pressure. Hence calculate the equilibrium constant. Formulate an equation for K_p in terms of the degree of dissociation.

8. Nitric oxide, NO, and carbon monoxide, CO, are air pollutants emitted by car engines. One possible reaction for eliminating them is

$$2NO_{(g)} + 2\,CO_{(g)} \rightarrow N_{2(g)} + CO_{2(g)}$$

From the data in Appendix 1, estimate the equilibrium constant for the reaction at 298 K. In a series of measurements, the pressures of each component in the air were $p(N_2)= 0.78$ bar, $p(CO_2) = 4 \times 10^{-4}$ bar, $p(NO) = 5 \times 10^{-7}$ bar, and $p(CO) = 5 \times 10^{-5}$ bar. In what direction would the reaction proceed under these conditions? Suggest why this may not be a practical method of removing NO from car emissions.

9. At 2500 K and 1 bar pressure, carbon dioxide is 15% dissociated into CO and oxygen. Calculate the equilibrium constant in terms of partial pressures and the standard Gibbs free energy change at this temperature for the dissociation. What would be the effect of changing the pressure?

10. The melting point of sulphur is 119 °C at 1 bar at which the density of the solid is 2.05 g cm^{-3} and that of the liquid is 1.811 g cm^{-3}. The enthalpy of fusion is 55.2 J g^{-1}. Calculate the melting point at 10 bar.

11. The normal boiling point of water is 100 °C and it has a vapour pressure of 92.51 torr at 50 °C. Calculate the entropy and enthalpy of vaporization.

12. For uranium hexafluoride, the vapour pressures of the liquid and solid phases are given by:

    solid: $\ln(p/\text{torr}) = 24.518 - 5895/(T/\text{K})$

    liquid: $\ln(p/\text{torr}) = 17.361 - 3479/(T/\text{K})$

    Calculate the enthalpies of evaporation, sublimation, and fusion.
    Calculate the pressure and temperature of the triple point.
    Can the liquid phase exist at 1 atmosphere pressure?

13. A mixture of benzene and cyclohexane containing 16.5 mol% cyclohexane boils at 78.3 °C at 1 atm pressure producing an equilibrium vapour containing 20.2 mol% cyclohexane. At this temperature, p° for benzene is 96.0 kPa and for cyclohexane is 94.2 kPa. Calculate the activity coefficients in the liquid phase.

    Given that the freezing point of pure benzene is 5.5 °C and its enthalpy of fusion is 10.51 kJ mol^{-1}, calculate the freezing point of the above mixture.

Further reading

Students who may have studied little physical chemistry before may find the first book a useful introduction to the basic ideas. Books 2–5 deal with thermodynamics in a way which extends the concepts and material covered in this primer. Thermodynamics is covered in virtually every undergraduate textbook on physical chemistry and some of those I find useful are listed in 7–11 (although this is a very subjective list!). 12 and 13 have an emphasis on biochemical thermodynamics. Finally, references 14–17 are books containing worked examples and problems. The easiest way to get to grips with physical chemistry is to do as many problems as possible. Good luck!

1. *Foundations of physical chemistry*, C. P. Lawrence, A. Rodger and R. G. Compton, OUP Primer Series No. 40, Oxford University Press 1996.
2. *Basic chemical thermodynamics* 4th Edn, E. B. Smith, Oxford University Press 1990.
3. *An introduction to the study of chemical thermodynamics*, 2nd Edn, D. H. Everett, Longman 1971.
4. *Chemical thermodynamics,* M. L. McGlashan, Academic Press 1979.
5. *Chemical thermodynamics : basic theory and methods,* I. M. Klotz and R. M. Rosenberg, John Wiley and Sons 1994.
6. *The Elements of physical chemistry*, 2nd edn, P. W. Atkins, Oxford University Press 1996.
7. *Physical chemistry*, 5th edn, P. W. Atkins, Oxford University Press 1994.
8. *Physical chemistry,* R. A. Alberty and R. J. Sibley, John Wiley and Sons 1971.
9. *Physical chemistry*, 3rd edn, G. W. Castellan, Addison-Wesley 1983.
10. *A textbook of physical chemistry,* A. W. Adamson, Academic Press 1979.
11. *Physical chemistry,* J. S. Winn, Harper Collins 1995.
12. *Physical chemistry with applications to biological systems*, 2nd edn, R. Chang, Collier Macmillan 1981.
13. *Physical chemistry with biological applications,* K. J. Laidler, Benjamin-Cummings 1978.
14. *Chemical thermodynamics—revision and worked examples,* H. P. Stadler, Royal Society of Chemistry 1989.
15. *Advanced physical chemistry calculations* H. E. Avery and D. J. Shaw, Butterworths 1971.
16. *Calculations in advanced physical chemistry*, 3rd edn, P. J. F. Griffiths and J. D. R. Thomas, Edward Arnold 1983.
17. *Beginning calculations in physical chemistry*, B. R. Johnson and S. K. Scott, Oxford University Press, 1997.

Index